BIG PLACES, BIG PLANS

Perspectives on Rural Policy and Planning

Series Editors:
Andrew Gilg
University of Exeter, UK
Keith Hoggart
King's College, London, UK
Henry Buller
Cheltenham College of Higher Education, UK
Owen Furuseth
University of North Carolina, USA
Mark Lapping
University of Southern Maine, USA

Other titles in the series

Geographies of Rural Cultures and Societies
Edited by Lewis Holloway and Moya Kneafsey
ISBN 0 7546 3571 6

Mapping the Rural Problem in the Baltic Countryside
Transition Processes in Rural Areas of Estonia, Latvia and Lithuania
Edited by Ilkka Alanen
ISBN 0 7546 3434 5

Power and Gender in European Rural Development
Edited by Henri Goverde, Henk de Haan and Mireia Baylina
ISBN 0 7546 4020 5

Young People in Rural Areas of Europe
Edited by Birgit Jentsch and Mark Shucksmith
ISBN 0 7546 3478 7

Multifunctional Agriculture
A New Paradigm for European Agriculture and Rural Development
Edited by Guido van Huylenbroeck and Guy Durand
ISBN 0 7546 3576 7

The Reform of the CAP and Rural Development in Southern Europe
Edited by Charalambos Kasimis and George Stathakis
ISBN 0 7546 3126 5

Big Places, Big Plans

Edited by

MARK B. LAPPING
University of Southern Maine

OWEN J. FURUSETH
University of North Carolina, Charlotte

Routledge
Taylor & Francis Group
LONDON AND NEW YORK

First published 2004 by Ashgate Publishing

Reissued 2018 by Routledge
2 Park Square, Milton Park, Abingdon, Oxon OX14 4RN
711 Third Avenue, New York, NY 10017, USA

Routledge is an imprint of the Taylor & Francis Group, an informa business

First issued in paperback 2018

A Library of Congress record exists under LC control number: 2004002029

ISBN 13: 978-0-815-38778-7 (hbk)
ISBN 13: 978-1-138-61895-4 (pbk)
ISBN 13: 978-1-351-16252-4 (ebk)

Contents

List of Figures

List of Tables

List of Contributors

Andrew Fisk, formerly with the Maine Land Use Regulatory Commission, is the Aquaculture Coordinator for the Maine Department of Marine Resources, Augusta, Maine, USA.

Owen J. Furuseth is Professor, Department of Geography and Earth Sciences, University of North Carolina at Charlotte, USA.

Hugh J. Gayler is Chair and Associate Professor, Department of Geography, Brock University, St. Catherines, Ontario, Canada.

Glenn R. Harris is Chair and Professor, Department of Environmental Studies, St. Lawrence University, Canton, New York, USA.

Michael G. Jarvis is Research Assistant for the Delaware State Energy and Transit Committee, a graduate student at the University of Delaware, and a graduate of St. Lawrence University, Canton, New York, USA.

Mark B. Lapping is Professor of Public Policy and Planning, Edmund S. Muskie School of Public Service, University of Southern Maine, Portland, Maine, USA.

Robert J. Mason is Director of the Environmental Studies Program and Associate Professor, Department of Geography and Urban Studies, Temple University, Philadelphia, Pennsylvania, USA.

Robert M. Sanford is Associate Professor of Environmental Science and Policy, University of Southern Maine, Gorham, Maine, USA.

Robert H. Twiss is Professor of Emeritus of Environmental Planning and Landscape Architecture, University of California, Berkeley, California, USA.

Jaap Vos is Assistant Professor, Department of Urban and Regional Planning, Florida Atlantic University, Ft. Lauderdale, Florida, USA.

Acknowledgements

This book could not have come together without the technical skills and patience of a number of colleagues at the University of North Carolina at Charlotte. We gratefully recognize Gary Addington, Carolyn Aguiar, Patrick Jones, Lisa Newman, Michelle McHugh, and Norma Redmond for their contributions to this volume. Finally, thanks to The Department of Geography and Earth Sciences for providing a collegial base for our work.

Mark B. Lapping
Owen J. Furuseth

To our families for their boundless support,
and to the staff and physicians at
The Southern Maine Dialysis Center

Chapter 1

Introduction

Mark B. Lapping and Owen J. Furuseth

With origins in the late 1960s, a 'quiet revolution' in land use planning and control has taken hold across North America for better than four decades. At least that is how it looks to numerous commentators and most especially to Fred Bosselman and David Callies who, in a report they authored for the President's Council on Environmental Quality published in 1961, actually coined the phrase the 'quiet revolution in land use control'. Some, like Frank Popper, saw the 'quiet revolution' as a manifestation of the 'arrival' of the environmental movement as a distinct and powerful force in national and local political life. Responding to numerous examples of environmental despoliation and on-going threats to large-scale ecosystems, the revolution prompted governments at several levels to attempt to protect critical areas and vulnerable environmental resources. Popper's view was largely coincident with that of Robert Healy as outlined in his 1976 classic, *Land Use and the States*. More recently Jerry Weitz has argued that the quiet revolution has matured into something more akin to a permanent revolution in that it has now been transformed into the 'Smart Growth' movement. But a few others, like Michael Heiman, have seen events differently. He maintained that the development of land use policy in the latter part of the twentieth century has been little more than the inevitable consequence of class conflict in which state authority was manipulated to protect elite patterns of land use and resource consumption.

No matter their views on the nature of the 'revolution', or the lack thereof, analysts and planners tend to agree that a certain amount of continuity exists between initial attempts at large-scale regional land use planning to protect endangered areas and ecosystems and the more contemporary impetus to reign-in sprawl across the landscape, the 'Smart Growth', movement. Aspects of continuity include the promotion of strategies to manage growth, recognition that projects of regional scale and impact can too rarely be effectively addressed by traditional local planning regimes, and that areas of critical concern-places of unique and complex environmental characteristics-often require the establishment of new planning systems and jurisdictions which superseded local planning prerogatives in the name of 'the public interest'. Not surprisingly, rural areas of significant natural and aesthetic character, and rural areas under substantial urban development pressure, have been and remain the focus of many of the revolution's planning schemes. Invariably these are large areas, big places if you will, which have become subject to big and complex plans.

Many of the big plans for these big places arouse out of their designation as 'critical areas'. They were defined that way not only by governments but also by interest groups, many with a considerable history of involvement in the regions and areas in question. Here the threat to environmental quality has been seen as severe and on-going, with environmental resilience decreased or compromised. More often than not, it was deemed insufficient merely to adopt a new attitude toward such places. Instead, it was necessary to develop entirely new planning and management systems and paradigms for such big places. This rarely happened without sustained controversy, contest, and even resistance on the part of local people and governments. Revolutions are, after all, messy, rarely smooth and almost never without conflict.

In each of the case studies contained in this volume, conflict and contest remains a sustained theme and 'part and parcel' of the planning process. This is a reflection of the different needs and aspirations of those most central to and involved in rural regional planning. It is also the consequence of long-standing class antagonisms and the shifting priorities of resource-based firms as the very notion of resource utilization changes from traditional uses and patterns, like forestry, small-scale agriculture and mining, to tourism, corporate farming, and real estate development. The backdrop and subtext remains that of environmental protection and how decisions will be made about the control and use of land.

Some tend to see all such big plans for big places as fundamentally anti-democratic in character and largely dismissive of indigenous knowledge and understanding. They argue that local planning is the most representative and the most responsible. This is certainly the view of James Scott (1998) in his wonderfully incisive *See Like a State: How Certain Schemes to Improve the Human Condition Have Failed*. Yet such a view can overlook the fact that in many cases 'local' often means that only certain long-entrenched and traditional elites will determine what happens in a place. Equity and distributional justice problems persist in both local planning regimes and in the extra-territorial governance patterns which have emerged to deal with the problems of these large rural regions.

The latter tendency in planning regimes, certainly a core artifact of the 'quiet revolution', has been justified by the insights of Gunderson, Holling and Light (1995) amongst others, who have argued that contemporary environmental problems require new institutional frameworks for planning, management and governance because those in existence are fundamentally incapable of dealing with emerging crises, 'surprises', and new patterns of resource use. In other words, existing resource planning systems are reluctant to change and adapt and too rarely can see change within the planning environment.

Much of the new thinking about planning governance often reflects a more robust, nuanced and sophisticated understanding of ecology and ecosystem resilience. This can be seen in the types of approaches to and strategies for regional planning reflected in the plans created for big rural places. Some were state or provincial level programs. Others reflected a more intrusive federal role than was traditionally the case. And still others attempt to strike a balance between local and extra-territorial and private and public interests. In only the rarest of these cases has the planning regime settled into a routine status. Conflict and challenge often over the very legitimacy of planning and state intervention continues and remains the norm.

Each of the chapters herein contained reflects the contemporary challenge of environmental and land use planning in large and substantial rural regions where changes in resource use, ownership and tenure patterns, and the restructuring of economies and societies has been taking place for several decades. In a few cases, such as those of the Adirondacks and northern Maine, decline is nearly a century old as the major contour of life. In some regions, like the Niagara Fruit Belt, the Everglades, and Vermont, the changing nature of agriculture and urbanization have come to mold much of the debate. The impact of recreation and tourism, second-home development, and severe stress upon the environment, most especially water use and quality, flavor all of these case studies to a great degree, but most especially in the Everglades, the Pinelands of New Jersey, the Adirondacks, and the Lake Tahoe basin. What remains common to all of these case studies is the fact that each region in question is a big rural area which is managed by a big and complex planning system. Descriptive and analytical studies such as these define much of the nature of contemporary planning in large rural regions across the North American continent.

References

Bosselman, Fred and Callies, David (1973), *The Quiet Revolution in Land Use Control*, US Government Pinting Office, Washington, D.C.

Gunderson, L.H., Holling, H.C., and Light, S.S. (1995), *Barriers and Bridges to the Renewal of Ecosystems and Institutions*, Columbia University Press, New York.

Healy, Robert (1976), *Land Use and the States*, Johns Hopkins University Press, Baltimore.

Heiman, Michael K. (1988), *The Quiet Evolution: Power, Planning and Profits in New York State*, Praeger, New York.

Popper, Frank (1981), *The Politics of Land-Use Reform*, University of Wisconsin Press, Madison.

Scott, James C. (1998), *Seeing Like a State: How Schemes to Improve the Human Condition Have Failed*, Yale University Press, New Haven.

Weitz, Jerry (1999), 'From Quiet Revolution to Smart Growth: State Growth Management Programs, 1960 - 1999,' *Journal of Planning Literature*, Vol. 14, pp. 268-337.

Chapter 2

The Beckoning Country: Act 200, Act 250 and Regional Planning in Vermont

Robert M. Sanford and Mark B. Lapping

Introduction

Known for its green image, picturesque villages, rolling farmland, maple sugar, Fall foliage, and Winter snow, Vermont is also recognized for its robust and muscular environmental regulations. The attention to environmental quality is immediately apparent by the absence of billboards along its highways. This small, rural state of 9,609 square miles has a total population of approximately 600,000. Its largest city, Burlington, has around 40,000 people and only 66 towns have populations that exceed 2,500. Despite, or perhaps because of its small size, Vermont has progressively experimented with land use regulation and planning. Like many states that experienced rapid growth and development in the post-World War II era, Vermont gained valuable planning experience in the war effort. This experience, although primarily in the allocation of resources and the managed production of goods, indicated the role planning could play in guiding growth and land use. But state and regional planning didn't happen overnight. It took several decades for the state and local communities to implement comprehensive plans. Regional planning commissions did not become effective until the mid-1970s. The State Planning Office remains a small department with links to economic development. Historically, there had been little incentive for regional or statewide planning.

Vermont's relatively small population had been expanding at an average of only 2 per cent a year from 1860 to 1960. This slow growth rate was easily accommodated and the state's official policy continued to encourage growth, notably second-home conversion of abandoned farms (Albers, 2000). State planners actively encouraged winter sport development and sought expansion of the state's infrastructure and services to encourage growth (Vermont Department of Housing and Community Affairs, 1999). By 1960, Vermont lost its ability to boast of having more cows than people, reflecting both population growth and a decline in the relative importance of the dairy industry (Byers and Wilson, 1983). Vermont's population growth was 12 per cent during the 1960s and 13 per cent in the 1970s. The growth in these two decades equaled the growth over the previous 135 years. Between 1980 and 1990, the state's total population increased to 563,000, a climb of 10.2 per cent. The 2000 census showed a decrease in the rate of growth; Vermont's population was at 608,827 (8.03 per cent growth from 1990).

The impressive growth of Vermont's permanent population has corresponded with increases in seasonal and transient population in association with a recreation and tourism boom that began during the affluent 1960s. The completion of the interstate highway system made Vermont much more accessible to the millions of urban dwellers of the metropolitan regions of New York, Boston, Hartford, and Montreal. Greater affluence, more leisure time and the ever-increasing popularity of skiing were contributing factors to the recreation-oriented boom that Vermont was experiencing. The state sought such growth, portraying Vermont as the 'Beckoning Country' in its marketing campaigns.

Growth and development in Vermont, as elsewhere, had some profound consequences. One was the increased demand for land. The heightened demand drove the price of land upward, particularly in fast-growing vacation resort areas where land was being converted to non-agricultural uses at a rapid pace.[1] Rapid development also created greater public service costs which precipitated an increase in tax rates. Higher taxes sometimes resulted in forced sales of land. Other problems occurred when land was hastily developed. Basic improvements, such as sewer and water systems, were often totally inadequate and barely addressed in local development review. The soils and topography of much of Vermont are not suitable for development and require meticulous planning if they are to be developed at all. Rapid and unplanned development of the 1960s was particularly troublesome and created substantial environmental harm at a time when there was an emerging environmental consciousness throughout the United States. The beautiful mountainous landscapes and clear, swift-flowing streams, which served as an important amenity attraction were being degraded and abused by inappropriate development.

In the early 1960s, a number of prominent Vermonters, most notably Governor Hoff became concerned. In 1963, when Hoff took office, planning in Vermont was handled by several employees in a single state office and a scattering of local planners among some of the larger towns (Vermont Department of Housing and Community Affairs, 1999, p.4). Early in his first year of administration Hoff established the Central Planning Office, which was to coordinate state agency policies and the various planning activities at state, regional, and local levels. By 1965, regional planning commissions were authorized, with eight in existence in 1968 (Vermont Department of Housing and Community Affairs, 1999). County government had never been strong in Vermont and the regional commissions were seen by some as a way of gaining influence over state and federal agencies. Others, seeing dark clouds, were concerned about increasing government regulation and what they perceived to be a weakening of the commitment to private property rights and laissez-faire attitudes (McClaughry, 1975).

Land-use control mechanisms were unpopular and viewed as complex and time-consuming by many town and state officials as well as by a significant portion of the population. Hoff remarked, 'the word *planning* was anathema to most Vermonters', but also noted 'if I could have taken the legislature to New Jersey for a weekend, I could have come back and gotten all of my legislation through' (Vermont Department of Housing and Community Affairs, 1999, p. 5). Comprehensive land use regulation was an alien concept to many local governments in spite of the town and regional planning authority that had been provided by the state. Unlike most states, every square inch of land in Vermont has been divided into townships that serve as important administrative units. However, a few 'gores' still exist as unintended remnants of this task where surveyed boundaries failed to match geography, leaving tiny unassigned areas. Towns function in a way similar to counties in many other states. County government is still almost nonexistent in Vermont. Local government resources were minimal; towns did not generally have civil engineers, sanitarians, planners, or other land use professionals. Consequently, much of Vermont's rural landscape remained vulnerable to the whims of the developer even during the decade of the 1960s when Vermont first began to experience tremendous development pressure from the recreation and tourism industry (Schmidt, 1995).

There was particular concern about the course development of the state would take as a result of completion of the interstate highway link to Vermont. In 1968 the General Assembly acted to protect 'scenic values' by banning billboards and other off-premise signs and to control the location and operation of junkyards (Byers and Wilson, 1983). This unusual step helped to preserve the visual quality of Vermont when viewed from its main roads, although it did not appreciably reduce new development. In fact, it likely encouraged the sprawl of second homes, so-called 'trophy houses' capturing pristine views of the countryside.

The role of planning was strengthened in 1968 when the legislature created the Vermont Planning Council, an entity advocated by Governor Hoff. The Council issued a report, *Vision and Choice* (1968), articulating the importance of growth management through setting goals and objectives (Wilson, 1999). The Central Planning Office became the staff of the Council. The Council was to coordinate planning activities and establish broad goals and policies designed to coordinate planning activities at all levels of government. These initiatives laid the foundation for the more specific legislation adopted in the 1970s.

In the summer of 1968 a subsidiary of International Paper Company proposed a large second-home development on 20,000 acres of land in southern Vermont. Widespread news coverage of this project coupled with existing problems associated with hasty, ill-conceived resort subdivisions propelled the 'development crisis' from a localized issue in some southern towns to a state-wide problem. Attention focused on second-home and recreational developments, which were being built for the enjoyment of out-of-staters. Many residents became emotionally charged and quite concerned that the quality of life in Vermont was about to be degraded. Deane C. Davis became governor in January of 1969 and, after viewing some of the developmental problems in southern Vermont, personally intervened by requesting a voluntary halt to the International Paper project. Governor Davis instructed the Property Tax Division to report all land transfers of over 100 acres to his office so that the administration would be aware of where other large-scale development was likely to take place. In addition, his administration formed a 'development technical advisory team' to assist local officials in coping with problems such as water pollution, traffic safety, soil erosion, and other public planning issues. This measure was taken to provide local government the expertise needed to deal with these vexing land use problems. This action provided immediate attention to the most troubled parts of the state (Vermont Natural Resources Council, 1989a).

In May of 1969, Governor Davis created a special Governor's Commission on Environmental Control, chaired by Representative Arthur Gibb, a key member of the House Natural Resources Committee. The Governor directed the Commission to consider existing reports on the state's environment, and hear the views of qualified experts on environmental issues in an effort to strike a balance between continued economic growth and high standards of environmental quality. The Commission was intentionally composed of people representing a broad spectrum of interests. In January 1970, after meeting regularly for several months, the Commission made a number of recommendations, which formed the basis of the body of environmental legislation passed in the 1970 General Assembly, including Act 250.

The Commission identified the sparsely populated towns and the mountainous regions of Vermont as areas in need of special state protection. Its recommendations about how to provide this protection formed the skeleton of Act 250. The Commission advocated a state development permit process, which would review specified impacts of certain types of developments. A framework of state regulations and a statewide land use plan guiding land and resource utilization were considered necessary components of the state land development control program. The landmark achievement of the 1970 session was the passage of the Land Use and Development Law, Act 250.

Although the bill, which was adopted as Act 250 was based largely on Environmental Control Commission recommendations, the final product incorporated considerable compromise and included input from several administration officials, legislators, and citizens. The bill that emerged from the legislature created a permit process designed to utilize citizen review boards rather than state agency officials. The bill provided for regional administration of the state law and included local participation. Local participation was and continues to be a critical factor in assuring public acceptance of the process. The decision to locate primary review responsibility in regional citizen commissions gives the process a grassroots responsiveness and safeguards the system from charges of bureaucratic insensitivity to local issues. In addition to the permitting process, Act 250 called for a three-stage planning process to achieve a state plan.

The Act 250 State Plan and How it Withered on the Vine

Section 6041 of Act 250 called for the Environmental board to adopt an Interim Land Capability plan that would establish physical limitations for 'generalized' land use areas. This Interim Plan was adopted in 1972. Act 250 also required a Capability and Development Plan designed to guide an efficient economic development of the state (10 VSA Section 6042) that would be constant with the Interim Land Capability Plan and that would be used along with the ten criteria in reviewing projects (Section 6086(a)(9)). (see Figure 2.1)

A third plan, showing land use, was to consist of a map and statements of present and prospective land uses based on the capability and development plan, and was to be derived from previous two plans. In January of 1973, the capability and development plan draft was approved by both outgoing and incoming governors and

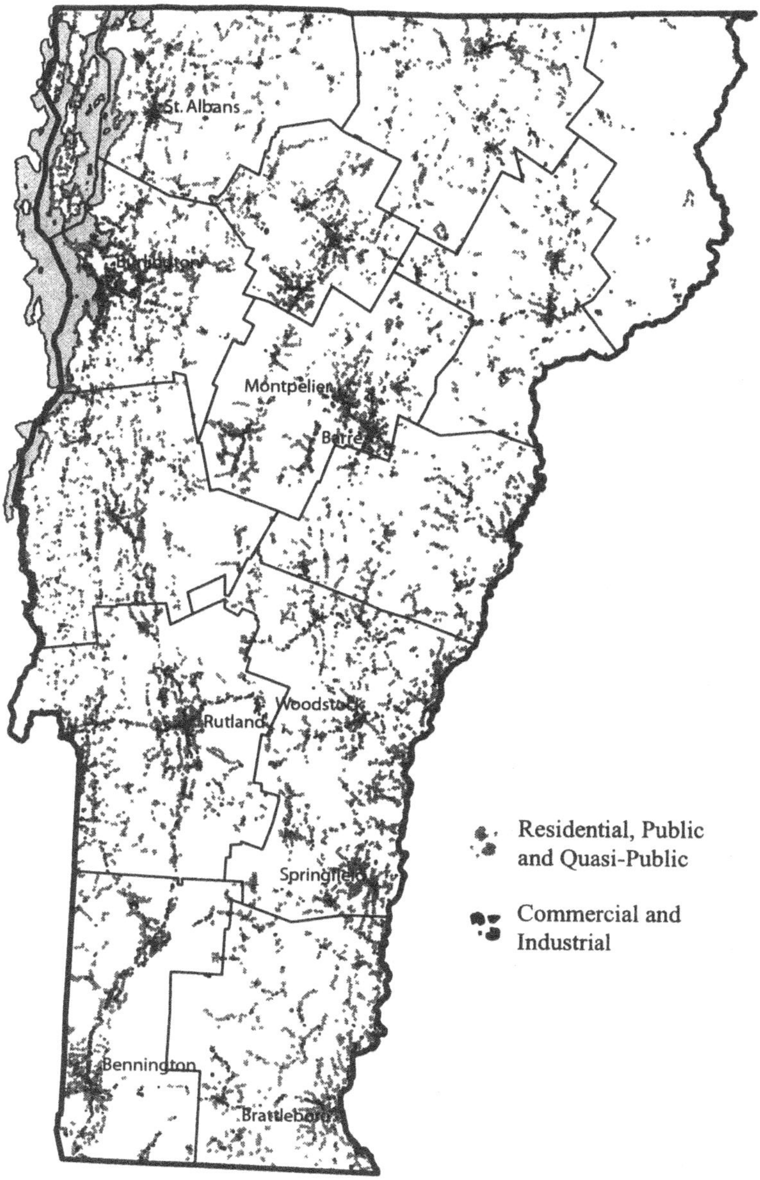

Figure 2.1 Interim Land Capability Plan

was finally adopted by the General Assembly – Vermont's legislative body – on the last day of the 1973 session. The legislation that was ultimately adopted included policy statements for use as guidelines by state agencies and local planning bodies and to provide the basis for the land use plan.

The land use plan was to contain a map and statements of present and prospective land use which determine broad categories of the proper use of lands (Brooks et al., 1997, p. XI-5). The plan contemplated both regulatory and non-regulatory approaches such as conservation easements, tax incentives and land acquisition. Language in the plan specified classifications of land into urban, village, rural, roadside and shoreline, and natural resource conservation areas. Former Environmental Board chairman Leonard Wilson describes the accompanying maps as rather hastily assembled, poorly printed and difficult to read (Wilson, 1999).

The merits of the land use plan were debated at eight regional hearings. Some Vermonters viewed the land use plan as tantamount to state-administered zoning. Opposition was also strong among land developers and various landowner organizations, who resisted attempts at top-down planning. Disagreements also emerged among state agencies concerning objectives of the plans, quality of the plan (including ecological validity) and on how the final versions should be worded. The legislature's land use study committee even included a minority opinion calling the majority report a 'unique medieval system' (cited in McClaughry, 1975, p. 13). The controversy and disagreement at the state level did not engender public support. In the end, these additions to Vermont's Environmental Control law did not have the broad-based backing that the original Act 250 legislation had received. Moreover, the legislature did not appear to be favorably disposed to any increase in state regulation because of the state's economic situation and the negative impact the energy (oil embargo) crisis had on the state's tourism and recreation sector.

The legislative draft of the land use plan never made it to the floor of the House in the 1974 session. A reformulated plan was introduced the following year but it too died on the vine, and 'Vermont forfeited a golden opportunity to guide its growth on a state-wide level with a heightened awareness of the ecology of the state' (Brooks 1997, p. XI-6). After the statewide land use planning effort failed, the rest of Act 250 continued in effect, receiving various amendments as legislators attempted to refine and improve the process. In 1983, Act 250 was amended to no longer require a state land use plan. However, the original intent of Act 250 remains intact and the Act itself continues to enjoy a broad-based support, enabling it to weather periodic legislative attempts at disembowelment (notably, the 1994-95 legislative year, in which several board members were ousted).

Although there is no statewide plan, individual state agencies have their own 'sector' plans that can be used in Act 250 decisions. These plans deal with specific resource categories such as agricultural soils and do not represent a comprehensive attempt at statewide planning. Instead, Act 250 has emphasized the role of regional and local plans as well as specific, project-based master plans (Brooks, 1997). Act 250 reviews projects on a case-by-case basis but the individualized nature of the process provides important degrees of flexibility in addressing cumulative impact review for growth management and in encouraging master planning.

Act 250 Administration

Act 250 is administered through nine district environmental commissions that receive and consider applications for development and subdivision permits. Each commission has its own defined area but can review projects in other districts if assigned by the state-level Environmental Board. The three commission members (and two alternates) for each district are appointed by the Governor and receive no compensation other than reimbursement for some expenses. They are chosen for their general familiarity with local conditions, and their civic-mindedness. Although appointees may represent a diversity of occupations, most do not have special training or expertise in environmental or land use planning, thus it is a true citizen board rather than a 'science court' or other specialized body. The idea is that environmental impacts and growth management issues, if properly articulated, are comprehensible to non-specialists and that citizens should have input as to how growth occurs and where. This is the heart of Act 250 (Sanford and Stroud, 1997). The district commissions encourage input through a public forum in which adjoining property owners, local planning boards, municipalities, other groups and individuals participate as 'parties.' Some parties receive automatic standing to participate (adjoining neighbors and municipal entities), others are granted standing if they can establish a likely 'direct effect' on their own property or if they have special expertise.

The district commissioners must receive a documented application, take evidence, hear challenges and reach positive conclusions under ten environmental criteria. To help do this, each environmental district is staffed by a district coordinator and a clerk, some have one or more assistant coordinators, and an extra clerk; thus a commission may have a staff of two but seldom more. Act 250 applies to most large commercial and industrial projects, although only about a quarter of all development in Vermont is actually subject to Act 250 jurisdiction, most development is too small to qualify for review. This was a concern even among the early drafters of Act 250 legislation – a fear that it might be ignoring the incremental impacts that collectively outweigh the large development impacts that actually get reviewed (McClaughry, 1975; Sanford and Stroud, 1997).

Act 250 land use permits are generally required for residential construction of ten or more units and subdivisions of ten or more lots. While state and local government projects are included, exemptions apply to electric power generation and transmission facilities and to construction for agriculture and logging purposes below the elevation of 2,500 feet. The scope of Act 250 jurisdiction over projects above 2,500 feet in elevation is particularly broad because of the environmentally sensitive nature of areas above this elevation. The Act defines development above this elevation as any 'construction of improvements for commercial, industrial, or residential uses.' Review is required of all 'substantial changes' made to projects, which were initially subject to Act 250 jurisdiction. A change is defined as substantial if it will have a significant impact on any of the ten criteria of the Act. Permit amendments for material changes, which may be of less import than substantial changes, are required for previously permitted projects.

A permit is also required prior to selling, offering for sale, or commencing construction on any subdivision. For Act 250 purposes, subdivision is defined as the partitioning of land for the purposes of resale into ten or more lots, within a five-mile radius (subsequently amended to include any lot within an environmental district). If towns do not have zoning and subdivision regulations, then jurisdiction applied to the creation of six or more lots. Until 1987, Act 250 matched Vermont Department of Environmental Conservation (DEC) septic design regulations by only counting lots if their size was less than ten acres. Currently (2002), the DEC continues to exempt residential lots of ten or more acres in its septic system design approvals Act 249).

Pre-1987 Act 250 exemption of ten-acre lots resulted in the widespread creation of large-lot developments designed to avoid Act 250 jurisdiction and fueling sprawl (Hamilton 1986; Hamilton and Clark, 1986). Problems still exist from this parcellization of previous years, from the exclusion of subdivisions with less than ten lots, and from the continued exemption of ten-acre lots under Act 249. The result is an incentive for sprawl. The most important part of the Act 250 process is the requirement that approved projects satisfy all relevant criteria. The appropriate district commission must receive a documented application, take evidence, hear challenges and reach positive conclusions. The applicant has a 'burden of persuasion', which it meets by submitting information to show compliance with the requirements of each of the components of the ten criteria. The first criterion requires projects to not result in undue water and air pollution. Criteria two and three review the availability of water supply and the ability of an existing water supply to meet the needs of the proposed project. Soil erosion and soil stability are reviewed under criterion four.

Criterion five addresses traffic impacts. Criterion six requires that a project not place an unreasonable burden on the ability of the municipality to provide education services. Compliance with criterion seven requires that a project not cause an unreasonable burden on the ability of local government to provide municipal or government services. Given that government has an obligation of government to accommodate reasonable growth through the provision of highway, educational, and municipal services, project applications cannot be denied under these criteria. While this might appear to conflict with the applicant's duties for a burden of persuasion, the police powers of the commissions enable them to compel evidence or to impose conditions of approval. Thus, the incentive remains for the applicant to make a complete disclosure to meet the burden of persuasion.

Under criterion eight, the commission must make sure that the project does not have undue adverse effect on scenic or natural beauty, aesthetics, historic sites or rare irreplaceable natural areas. This criterion has a subcriterion (8A) addressing wildlife habitat and endangered species, although it is recognized in practice that wildlife has ecological as well as aesthetic value.

Criterion nine requires that a project be in conformance with a capability and development plan and land use plan (this is the land use plan that was never adopted, however, this reference still remains). This criterion has several subcriterion dealing with specific growth impacts on resources such as agricultural soils, forestry, and energy (subcriteria, 9A - 9L). The tenth and final criterion addresses conformance with municipal plans and capital programs adopted by local governments and regional planning commissions. This last criterion is potentially the most powerful due to the regulatory strength it accords town and regional plans, documents otherwise used only for advice or guidance. Under Act 250, a project that violates a town plan cannot be approved. Interpretation of local plans in the Act 250 process requires careful scrutiny. Zoning and other local land use regulations can also be looked at in the context of understanding the plan (Vermont Supreme Court, 1994). Effective July 1, 2001, Criterion 10 was amended to allow the board or district commission to consider town bylaws if the town plan is found 'to be ambiguous'.

These criteria have created a planning system that focuses on a project's likely performance relative to the integrity of environmental and community resources. In this, Act 250 represents the maturation of a type of planning approach that emphasizes performance standards rather than strict land use categorizations. This is consistent, to some extent, with the tradition of 'mixed uses' encountered in most Vermont towns and villages. It also requires local resource knowledge and local citizen involvement to be most effective. The location, nature and frequency of Act 250 hearings is a key issue in maintaining public involvement.

District commissions generally receive 20 to 90 applications a year. Usually, an applicant will receive a permit within 60 days (Sanford and Stroud, 1997). After an application is submitted, a hearing must be scheduled within 25 days and held within 40 days unless waived by mutual agreement of all potential parties. A permit may only be denied if the commission documents specifically show the project would be detrimental to the public health, safety or general welfare. The vast majority of permits are granted with conditions on one or more criteria.

Effectiveness of Act 250 in Regulating Growth

Since its passage in 1970, the Land Use and Development Law (Act 250), has succeeded in controlling a significant portion (though still a minority) of growth and development in the state. Over the years, Act 250 has become accepted by the general public and there are no apparent signs that it will be replaced or eliminated. One of the most significant aspects of Act 250 is that it utilizes citizen participation in the decision making process. Though not without its critics, it has gained the widespread support among the general population and among government officials as well. An important part of this acceptance is the role that citizens play in the decision-making process as commissioners and as 'parties'. Such participation has been crucial in helping Act 250 gain widespread, grass roots support (Bourdon, 1995). Because Act 250 applies to state and municipal projects involving more than ten acres, local citizens receive a direct say in government development decisions. However, in conservative echelons, Act 250 (and Act 200) have been attacked and subjected to periodic attempts in the legislature to disembowel land use regulation. Former Vermont state senator John McClaughry described the planning movement in Vermont (and nation-wide) as an attempt by the state to centralize all power over land, calling it 'the new feudalism' (McClaughry, 1975).[2] On the opposite side of the coin, attempts have been initiated to strengthen regulation, notably in response to the rise of corporate land speculation (Hamilton, 1986) and recognition of the need for growth management (Brooks, 1997).

There is disagreement among governmental officials, developers, and the general public as to whether Act 250 is the most effective way to protect the environment and preserve a healthy economy (Brooks, 1997). Supporters of the legislation cite several examples of positive impact including an improved economy, energy conservation, fish and wildlife preservation, aesthetics, and water resources. According to advocates, by protecting Vermont's natural resources, Act 250 also protects the significant revenues generated by tourism, residential recreation, 'green trade,' agriculture and forestry. These sectors generate over one and a half billion dollars in revenue for the state's economy each year. By preventing resource degradation, Act 250 saves or substantially reduces costs of remediation. Other positive economic impact features include tempering speculation and weeding out many ill-conceived large projects (Dafoe, 1992). Act 250 has had a positive impact in many other areas as well. For example, criterion 8(A) provides a regulatory mechanism for wildlife protection. This protection is one of the few areas of Act 250 review where some economic balancing is required. If a development or subdivision 'will destroy or significantly imperil' necessary wildlife habitat or endangered species, a District Commission is required to weigh the economic, social, cultural or recreational benefit to be derived from the development versus the economic, environmental or recreational benefit derived from the habitat or species. Another important consideration is whether all 'feasible and reasonable means of preventing or lessening the destruction or imperilment' will be implemented. In most cases, projects can be designed or redesigned to avoid undue adverse impacts on important resources (Vermont Natural Resources Council, 1989b). Ski areas and other large landholders generally must prepare a wildlife management plan to comply with this criterion.

Aesthetics has also become an important issue in growth management in Vermont under Act 250. In considering aesthetics, the review process addresses two essential questions: Will the project have any 'adverse' aesthetic impacts on the scenic quality of an area and, if so, whether those impacts will be considered 'undue' when taking into consideration the type of development project and its surroundings. In its analysis, the District Commission or the Environmental Board must determine if the proposed project is compatible with its surroundings in terms of its visibility, impact on open space, and visual sensitivity of the location. If there is adverse impact, then the next step is to determine if any of the following are true: Will the project violate any clear, written community standard? Would the average person find the project shocking or offensive? The district commission or board is held to be the equivalent of an 'average person' and thereby need not be drawn into reviewing petitions or determining local sentiments. Has the applicant failed to take reasonable steps to lessen adverse effects? If the answer is yes to any of the above questions, then the project will be considered 'unduly adverse' and must be denied. However, it is rare for a project to be denied on purely aesthetic grounds.

Often, the design of an adverse project is altered or the size of the development is reduced in order to comply with the criterion. Further, this approach, called 'the Quechee Analysis',[3] is capable of surviving legal challenges of vagueness or arbitrariness - the structured process allows the incorporation of methodical visual impact assessment techniques to obtain replicable results (Brooks, 1997). This concern with environmental aesthetics also reflects a long-standing Vermont policy tradition that saw the state adopt the nation's first beverage container law in the mid 1950s and several other measures to protect country roads (used as a model for federal legislation), and riparian environments as well as in the aforementioned highway sign law (Lapping, 1982).

By considering the overall change in the hydrology of a watershed, Act 250 provides an ecological perspective in protecting environmental resources that might fall between the cracks of other statutes or regulations. Act 250 allows consideration of 'big picture' items as well as the review of seemingly environmental minutia. However, despite the relative scrutiny of land development of all kinds under Act 250, the approval rate is quite high. These approvals occur through a process of negotiation and mitigation.[4] The result is a much improved land development process with projects that are much less detrimental to Vermont's beautiful mountain and valley landscapes (Vermont Environmental Board, 1995). The incorporation of citizen participation into the review process reflects Vermont's strong participatory democracy orientation-town meetings are still a zealously defended tradition. The perceived small size of the state and the sense of community held by its citizens contributes to the role of neighbors and town residents in the Act 250 process. Not only do citizens appear at Act 250 hearings as witnesses and parties, but citizens serve on commissions and boards, with only a skeletal staff. Even though the denial rate for permits is extremely low, only about 4 per cent, virtually all approved projects do have conditional requirements, based upon either formal findings or the applicant's acceptance of proposed 'minor' permit conditions when no hearing is held. Despite its many positive features, the Act cannot reach its full potential without a comprehensive land use plan. Absent such a plan, there is no corresponding planning to match the Act 250 case-by-case reviews. Another issue that affects planning and is associated with case-by-case review is the problem of addressing cumulative impacts. Cumulative impacts can escape review or mitigation until they involve a project that happens to come under Act 250 jurisdiction.

Very few projects are denied under Act 250. Approximately 95 per cent of Act 250 applicants get permits, usually within 90-120 days. Occasionally, a large project that is particularly controversial could get caught-up in a lengthy review appeals process but this is the exception rather than the rule. Yet such cases may receive an unusual amount of publicity, belying the true facts. Scrutiny of the decisions handed out by the Environmental Board shows that of the 14 cases appealed to the Vermont Supreme Court between 1990 and 2000, only one was reversed, proving that in the view of the court, the board was not subjective but following the letter of the law over 90 per cent of the time (Sanford and Stroud, 1997). One individual or corporation, town, and other parties have similar rights to stop a project only if they convince the review board that there is some legitimate or compelling reason why it should not proceed. A large part of the popularity and attractiveness of the Act 250 process is based on grass roots support and citizen participation. The power of an individual or family should be no less important than that of a developer or large corporation according to Dick McCormack (1994), former Act 250 district commission chairman and long-time chair of the Senate Natural Resources Committee. However, not all parties have equal footing in the process. Permitted parties, for example, unlike statutory parties, cannot appeal projects to the Vermont Supreme Court. Environmental watchdog groups, such as the Vermont Natural Resources Council, claim that this significantly hampers their effectiveness (Vermont Natural Resources Council, 1999).

Despite charges and counter charges, complaints about Act 250 continue to rise, although the statistics favor the Act's defenders who say Act 250 is an undeserving and convenient target for developers frustrated with a recessive economy (Cannon, 1993; Dillon 1994). Act 250 does not stop the vast majority of developments. Many projects are improved through the process, by making design more effective, adding stream buffers or other considerations that make the use of a site more successful. Most permits are issued without hearings and within two months. Act 250 can be generally considered as unhostile to development. It monitors growth and development and through mitigation improves the developments that are being proposed (Pfeiffer and Dillon, 1994). Further, other, non-regulated environmental impacts may be much greater than those of Act 250 (Olinger, 1993). Land speculation and subdivisions designed to avoid Act 250 review can cause some land use problems (Sanford and Stroud, 1997, pp. 251-252):

1. Distorted lots: Lot shapes are distorted to allow developers to maximize road frontage or shore frontage and still be over ten acres, escaping review of water supply and septic systems by the other state agency that approves subdivisions, the Department of Environmental Conservation. Many parcels are carved up into 'spaghetti' or 'bowling alley' lots because of the long, narrow design that avoids shared roads. Without a regulatory or planning incentive, subdividers may have little motivation to pay attention to the natural lay of the land.
2. Reduction of the working landscape: Unregulated subdivisions may engender the piecemeal removal of productive forest and agricultural land from use (Lapping, 1978). Widely dispersed ownership makes commercial timber management and harvesting impractical. Dispersed parcels also inhibit agriculture by reducing critical mass and fostering "right-to-farm" problems (Lapping, et. al., 1983).
3. Reduction of the commons: Extensive land subdivision also removes and reduces the amount of land available for public recreation such as cross-country skiing and hunting.
4. Habitat fragmentation: Wildlife habitats such as "deeryards" (deer wintering areas) are made increasingly vulnerable as more and more land is subdivided into small parcels that may someday become house sites. Wildlife corridors may be blocked.
5. Health risk: Most towns in Vermont lack the ability to adequately review septic systems and water supplies, let alone other potential sources of impact reviewed under Act 250 or Department of Environmental Conservation regulations. Parcels divided into nine or fewer lots all over ten acres in size may receive no substantive review at all.
6. Inconsistent review: A mixed message is being sent to the public. The potential for environmental impact is not necessarily related to the size of a lot. Further, an insensitive five-lot subdivision may have more adverse environmental impacts than a professionally designed and developed subdivision of ten or more lots (Hamilton and Clark, 1986).
7. Lack of growth management planning: Subdivisions planned to avoid jurisdiction can represent poor land use. Towns can become overwhelmed with individual projects and land the time for proper, anticipatory planning. Infrastructure can become committed, extending services to areas not suited for development or not in the best interest of smart growth strategies. Proper layout and design may be ignored in their attempts to avoid the Act 250 review process (Hensel, 1996). Daniels and Lapping (1984) and Hamilton (1986) note the large amount of land sales and development activity that bypassed the provisions of Act 250, receiving little if any regulatory review.

Act 250 and Project Master Plans

The statute is not necessarily conducive to planning. Many developers, particularly in the competitive resort industry, are reluctant to submit a master plan for review since it alerts potential opponents to the project and to ultimate plans for site build-out. The developer may be concerned that the master plan suggests far greater impacts than the project currently under review thereby leaving the project vulnerable as a potential edge to get at these future impacts. The Environmental Board is not set up to manage property or deal with planning, however, the board's Rule 21 allows for partial findings of fact, which formed the basis for developing master plan reviews.[5] In addition to managing potential environmental impacts, there have been some interesting efforts to provide affordable housing through the review of master plans. These efforts led to the Board's adopting a master plan policy (see below). Stratton Corporation and Intrawest Stratton Development Corporation ('Stratton') created an affordable housing agreement with the Vermont Housing and Conservation Board in accordance with Land Use Permit #2W0519-17 (Valley View Project Affordable Housing Obligation) issued by the Vermont District #2 Environmental Commission pursuant to 10 V.S.A. Chapter 151 (Act 250). The permit approved construction and built-out of the Stratton Master Plan (1,370 housing units, clubhouses and associated ski facilities) in the towns of Stratton and Winhall. The commission concluded that Stratton had an obligation under Criteria 9(A) *Impact of growth* and 10 *Conformance with local and regional plans*, to ensure perpetually affordable housing to help offset some of the impacts of Stratton's growth. (Stratton Mountain Affordable Housing Agreement, Act 250 Permit #2W0519-17, 2000). The commission recognized that the growth from the project had created a need for workers. The increased housing demand exacerbated the existing low levels of affordable housing. Okemo Mountain Inc. also constructed affordable housing units via a similar agreement in the mid 1990s (Hensel, 1996; Land Use Permit #2S0351 and various amendments).

To compensate for the problem in obtaining master plans for review, the Environmental Board adopted a specific master plan policy (adopted in 1998, revised in 2000):

> The objective of the master permit policy and procedure, pursuant to Board Rule 21, is to provide guidance and greater predictability to the applicant and all parties in the review of complex development projects. Pursuant to Board Rule 21, the applicant may seek permission from the district environmental commission or the board on appeal to proceed with review under specific criteria of the Act in order to gain a greater degree of assurance that future development projects may be approved on a proposed development tract. This procedure will allow for greater efficiency in the environmental review process and therefore avoid unnecessary and unreasonable costs to the applicant and parties.
>
> Before the district commission or the board can grant a master permit for a master plan project, it must be able to make positive findings of fact and conclusions of law demonstrating compliance with all ten criteria of Act 250 for those aspects of the project seeking construction approval. For most, but not all of these issues, the burden of proof lies with the applicant pursuant to 10 VSA Section 6088. In order to receive a master permit with partial findings of fact, an applicant must carefully define the

intended scope of the development and sufficiently quantify project impacts.

In order for the district commission or the board to issue a permit authorizing construction approval for a certain phase of a master plan, positive findings of fact and conclusions of law must be made under all criteria for that particular phase. In many instances, a permit may be granted for a smaller portion of the total project (including infrastructure) with partial findings of fact for the remainder of the project under the relevant criteria. These partial findings of fact will provide guidance and greater predictability to the applicant in preparing final plans for the project or for subsequent phases. Partial findings of fact are generally issued for a period of five years since this represents a reasonable planning period within which potential impacts under the relevant criteria can be ascertained. Prior to expiration, partial findings of fact may be renewed and updated as necessary. In each Regional Office, District Environmental Coordinators are available to assist applicants in proceeding under 10 V.S.A. Section 6086(b) and Rule 21 which provides the legal framework for the full implementation of this policy and procedure. In most instances, the initial review of a master plan application will focus on the project's scale, location and impacts under the so-called "natural resource" criteria of the Act, including but not limited to, Criteria 1(A), 1(D), 1(E), 1(G), 8, 8a, 9(B), 9(C), 9(D) and 9(E).

While this master plan procedure does not represent planning by a regional or state entity, it is significant in that, if the state asks master planning of its large developers, it is reasonable in turn to ask the state about its own planning efforts. Further, many of the large resorts, such as Stratton and Okemo, straddle several towns and may involve more than one regional planning commission. Given the scale of Vermont itself, such developments are seen as regional in and of themselves. Master plan procedures reinforce proponents of planning and the general culture of accountability for growth management.

Other State and Regional Planning Efforts

The Land Gains Tax of 1973 was enacted to discourage short-term speculation in land by taxing land sold within six years of acquisition. A 1987 amendment further targeted land speculation occurring within four months of ownership. The Housing and Conservation Trust Fund (1987) provided $130 million for affordable housing, farm conservation and open space planning. The 1993-1995 Growth Centers Pilot Project led to a HUD Consolidated Plan (1995), a Vermont Highways Standard (1997) promoting greater sensitivity to adjacent environments, and an Access management Program (1998) to ensure curb cuts on state highways are consistent with state land use goals as well as with state, regional, and approved local municipal plans. The pilot project also contributed to property tax reform in 1998 that, although focused on educational resources, has tax-sharing provisions that affect land use decisions. The 1994 Vermont Downtown Program and the 1998 Downtown Development Act both serve to concentrate resources in urban and village centers. An Interstate Interchange Policy was put into effect in 1999 in recognition of highway interchanges as precursors of development. The policy seeks to ensure that development (or conservation of environmental resources) at interchanges is consistent with state land use goals.

In 1991, the Lake Champlain Basin Program was created to develop a comprehensive plan for the basin region, covering about a third of the state. The plan, *Opportunities for Action: An Evolving Plan for the Future of Lake Champlain*, was completed in 1996 (Lake Champlain Basin Program, 1996). Despite the plan, Lake Champlain receives a significant amount of nutrient load (18 per cent) from urban runoff and sprawl has caused area watersheds to exceed the target of no more than ten per cent impervious surface (Courtney, 1999). Poor growth management has fueled the decline of small farms and the consolidation of cows into 'massive, productive agricultural factories,' making the largest single contribution to pollutant loading in Lake Champlain (Lindner, 1999).

The Vermont Consolidated Plan (Vermont Department of Housing and Community Affairs, 1997) addresses the use of federal funds in community development planning, linking historic preservation groups, industry and business, and other partners in guiding growth. The proven provides encouragement for regional planning as well as for downtown redevelopment. The Vermont Housing Council commissioned a Vermont Housing Needs Assessment (1998) that showed the increase in homeless families and the shortage of affordable housing, despite the rosy economic outlook for Vermont. Housing is addressed in regional plans but the lack of suitable housing shows the ineffectiveness of these plans and state-level initiatives. The Vermont Housing Council is attempting to coordinate more effective planning, including use of the HUD Consolidated Plan to improve country, regional, and state agency and nonprofit group coordination in addressing housing needs (Vermont Housing Council, 1999).

The Vermont Community Development Program (VCDP), funded by the federal Community Development Block Grant (CDBG), provides planning grants to give communities technical expertise and professional resources needed to test ideas, propose strategies, develop plans, establish policies and procedures, and conduct organizational activities. These grants assist in community development planning activities and preparing to implement projects in one or more of the program areas; housing, economic development, public facilities or public services. VCDP also provides implementation grants and interim financing.

The Vermont Agency of Commerce and Community Development's Strategic Plan (2000) contains strategies for growth management and planning, notably:

> Strategy 2.2 Enhance communities' ability to plan for and implement growth to support Vermont's traditional land use pattern.
>
> 2.2.1 Develop a strategic plan for smart growth and land use planning at the state, regional and local levels.
>
> 2.2.2 Continue to research and evaluate 'new urbanist' and 'smart growth' strategies for transfer to Vermont.
>
> 2.2.3 Continue highway interchange planning. Complete action plans for interstate interchanges.
>
> 2.2.4 Protect historic resources through funding and other protective measures such as Section 106 and Act 250. Add training opportunities for professional and citizen

planners: Section 106, Act 250, annual Division for Historic Preservation conference or municipal officers and town official's conferences.

2.2.5 Provide resources and technical assistance to local planning and zoning officials. Reprint planning manual, Chapter 117. Publish open space planning resource guide.

2.2.6 Maintain web page. Add interactive map; update existing information; add History of Planning, Planning Manual, Chapter 117, etc.

2.2.7 Promote new Certified Local Governments [CLGs] and continue technical assistance and funding program. Add two CLGs.

2.2.8 Disseminate model bylaws, planning techniques, and other resources among communities. Hold annual planning celebration.

State oversight of regional planning commissions has led to increased coordination among the commissions. For example, three regional planning commissions including one that straddles two states, covering parts of Vermont and New Hampshire convened a summit on affordable housing on March 28, 2001 (Two Rivers-Ottauquechee Regional Commission, Southern Windsor Country Regional Planning Commission and Upper Valley Lake Sunapee Regional Commission, 2001).

The Vermont Forum on Sprawl, funded by the Orton Foundation, has launched a major growth management initiative, providing publications, workshops, resources, and strategies for growth management in Vermont. Its overall strategy is to promote planning, policy, investment, incentives and regulations that reinforce growth management. Specific projects targeted by the Forum (1999) are:

1. Passage of the Downtown Bill in partnership with Downtown Coalition and State of Vermont.
2. State strategy to coordinate state public investment in accordance with smart growth principles.
3. Closure of the Ten Acre Loophole [residential lots over ten acres are generally excluded from Act 250 regulation].
4. State appropriations that support smart growth, including appropriations for downtowns, local planning, Housing and Conservation Board, and key transportation investments.
5. Support for maintaining the integrity of Act 250.
6. Work with individual state agencies and departments on anti-sprawl measures. Examples: Vermont. Economic Progress Council re: guidelines for tax incentives, Agency of Natural Resources water and sewer grant/loan criteria, pilot village waste water management.

The Forum advocates the following "Smart Growth Principles" (Vermont Forum on Sprawl, 1999):

1. Plan development so as to maintain the historic settlement pattern of compact village and urban centers separated by rural countryside.
2. Promote the health and vitality of Vermont communities through economic growth and residential growth that is targeted to compact, mixed use centers,

including resort centers, at a scale appropriate for the community and region.
3. Invest public funds consistent with the vision for Vermont communities under 1. and 2. above.
4. Enable choice in the mode of transportation available and insure that transportation options are integrated and consistent with land use objectives.
5. Protect and preserve the environmental quality and the important natural and historic features of Vermont, including natural areas, water resources, air quality, scenic resources, and historic sites and districts.
6. Encourage and strengthen agricultural and forest industries and minimize conflicts of development with these industries.
7. Provide for housing that meets the needs of a diversity of social and income groups in each Vermont community but especially in communities that are most rapidly-growing.
8. Support a diversity of viable business enterprises in downtowns and villages, including locally-owned businesses, and a diversity of agricultural and forestry enterprises in the countryside.
9. Balance growth and the location of growth with the availability of economic and efficient public utilities and services.
10. Accomplish goals and strategies through coalitions with stakeholders and engagement of the public in solutions for smart growth.

Act 200

In response to the rapid rate of growth in Vermont, Governor Madeleine M. Kunin initiated the Governor's Commission on Vermont's Future on September 12, 1987. The commission held hearings all around the state in the fall of 1987; a town meeting writ large. The commission issued a report in January of 1988 that articulated Vermont values and concerns. The report containedguidelines for planning, economic development, natural resources, agriculture, and affordable housing. The commission called for 'a coordinated, comprehensive planning process and policy framework ... to guide decisions by local governments, regional planning commissions and state agencies' (1988, p. 21). The report emphasized the importance of citizen participation 'at all levels of the planning process,' and the planning of a reasonable (matching ability to provide municipal facilities and services) rate of growth that 'maintain[s] the historic settlement pattern of compact village and urban centers separated by rural countryside' (1988, p. 21).

On May 19, 1988, Governor Madeleine Kunin, a Democratic advocate of planning, signed the Growth Management Act of 1988 (Act 200) into law. This Act was the result of those public meetings and was intended to improve planning at local, regional and state agency levels by providing incentives. Towns whose plans conformed to regional plans were eligible for increased funding and resources. The law also allowed communities to assess impact fees. Lack of funding as well as political opposition to planning has hampered the Act's effectiveness. By 1999, most (95 per cent) Vermonttowns had written town plans but only 55 per cent of the towns had obtained approval of their town plans by the regional planning commission. Although the approval process is not rigorous and despite the fact that the towns are represented on the regional commissions, there is simply not enough incentive for fuller participation. Still, the Act attempts to instill 'good planning' through the following key provisions (Vermont Department of Housing and Community Affairs, 1988):

1 Planning goals: Set up 32 broad, flexible planning goals to be followed.
2 Town plans: Towns that have conforming plans qualify for technical assistance, additional funding, and greater local influence over state actions.
3 Regional Planning: Each town became a member of its regional planning commission, which in turn, must ensure that its plans are "compatible" with approved town plans in the region.
4 State Agency Planning: All state agencies that make land use decisions must adopt plans and take actions consistent with the 32 state planning goals and the goals and plans developed by towns, regional planning commissions, and other state agencies.
5 Council of Regional Commissions: A council was created with representatives from each regional planning commission along with three state agency heads and two citizens appointed by the governor. The council is tasked with reviewing regional and state agency plans to ensure they are compatible with each other and with the 32 planning goals.
6 Geographic Information System (GIS): Active support of the development of GIS applications and technology for planning.
7 Housing and Conservation Trust Fund: A nine-member board administers the Trust Fund (intended at $22.5 million for the 1987-89 period and over $3 million per year subsequently) to preserve open land, protect historic and environmental resources, and promote affordable housing.
8 Property Transfer Tax: This tax was increased to 1.25 per cent to help fund the Trust Fund and other planning funds. The first $100,000 of primary residences and all working farms remains at the former rate of 0.5 per cent.
9 Impact Fees: Some communities had previously used Act 250 to levee impact fees. This provision clarified municipal authority for fees to offset municipal costs caused by development. The fees must be directly linked to the development's impacts and must be spent on the needed capital improvement.
10 Farm Programs: The Dairy Income Stabilization Program was established to provide immediate payments of $5,000 to dairy farmers who make more than 50 per cent of their income from farming. The Working Farm Tax Abatement Program was also set up. This program pays 95 per cent of the property taxes on open land and farm buildings for non-dairy farmers in the first year of the program operation and for all farmers in future years.

The planning goals for Vermont are broad policies that incorporate traditional Vermont values of community life, agricultural heritage, environmental quality and economic quality. These goals apply to state agencies, regionalplanning commissions and towns.

Vermont planning funds 'declined precipitously' shortly after the passage of Act 200 and remained low through the1990s (Vermont Department of Housing and Community Affairs 1999, p. 14). A Growth Centers Pilot Project was financed from 1993 to 1995 to promote the promise of growth centers articulated in Act 200 and seeking a streamlined permit review system in Act 250. Vermont continued to promote a personal, localized approach to planning. The 1998 Downtown Bill provided incentive for reinvestment in urban centers. The state itself determined to set a leadership role, relocating state environmental offices from the fringes to downtowns in Rutland, Springfield, and other urban settings. Governor Howard Dean even traveled to Wal-Mart headquarters in Arkansas to make the case for Wal-Mart to build in downtown Rutland and other downtowns rather than engage in its traditional tactic of fringe construction (Vermont Department of Housing and Community Affairs, 1999, p. 15). Wal-Mart did indeed build in the downtowns of Rutland and Bennington.

Discussion

After 30 years, Act 250 continues to be Vermont's landmark land use and development law. Despite several amendments and a few changes here and there, it continues to receive strong support from Vermonters. There are, however, several shortcomings that affect planning. The Act may lead some residents to be complacent, concluding that they need not worry about local regulations because Act 250 will provide the necessary local protection(Merrill, 1987, pp. 22-24). As a result, very few Vermont towns have established effective environmental review by conservation or planning commissions for natural resources. Further, most small land development projects are excluded from Act 250 review, leaving many towns ill prepared to deal with negative environmental impacts and poorly planned growth. This places a substantial responsibility for development approvals in the hands of individual towns, many of which lack adequate resources and expertise to meet this responsibility (Daniels and Lapping, 1984, pp. 502-508).

Growth management problems associated with commercial development, residential subdivision, and traffic are increasing, to the detrimental of the quality of life in Vermont (Brooks, 1997). Without the statewide land use component, Act 250 does not appear sufficient to manage rapid growth. A land use component would provide the state with a clearer sense of direction. Even though most towns and all of the regions have adopted plans, they are usually very general and often fail to address essential planning decisions. Without comprehensive local and regional plans, basic planning decisions are either not made or are left to the regulatory process.

Towns could improve growth management through exerting greater control over local development projects through the adoption of zoning bylaws, subdivision regulations, capital improvement plans and official maps. If towns were granted the authority by the General Assembly in the site plan reviews to more adequately address criteria such as drainage, soil erosion, and aesthetics, it could greatly enhance the quality of a project and also shorten the review process at the District level (Governor's Commission on Vermont's Future, 1988). Despite the early criticisms of statewide planning, Vermont has always been interested in local participation, even in a state land use law. Although it is doubtful the statewide plans could be resurrected, increased use of regional planning commissions in Act 250 and additional fine-tuning of Rule 21 could promote planning beyond the level of the individual site. Act 250 could be amended to drop the small development exclusions, strengthen cumulative impact reviews through master plans, and improve monitoring and compliance procedures. Or these tasks could be more fully incorporated into Act 200. Act 200 compliant towns could receive greater concessions under Act 250.

The sense of community in Vermont may contribute to growth management in the future. State government in Vermont is at a par in size and complexity to local government in a large metropolitan area. However, there are disparate small municipal governments in Vermont and to capitalize on the larger sense of Vermont community requires regional or state oversight of these smaller units. In general, politically fragmented urban areas tend to sprawl more, based on a national study of density trends (Fulton *et al.*, 2001). Ironically, Fulton *et al.* (2001, p. 14) found that density dropped more rapidly in metropolitan areas in states with legislation requiring local plans to submit comprehensive growth plans to a state agency for review. Fulton suggests that this may be because state agencies might lack the clout to ensure proper local plans and their implementation or it may be because an urban area closes off to growth in response to a lack of infrastructure (or a need to first improve it) as identified in a growth plan, edging out development into a rural area that has the infrastructure (or that is unaware that it lacks it).

Despite the lack of comprehensive large-scale planning, and because of the efficacy of Act 250, Vermont ranks well in terms of environmental laws and economic growth (Cannon, 1993). Improvements could propel Vermont to the top or near the top of the list among states with effective statewide comprehensive land use and development guidelines.

Klyza and Trombulak (1999) envision three potential Vermont landscapes in 2014: a Vermont based on current trends, with less farms and less wild lands; an overdeveloped Vermont that looks like much of southern New England; and a Vermont based on logical land use controls and growth management. Whether this final vision is achieved is anyone's guess.

Notes

1 The state has had a history of encouraging the conversion of abandoned farmlands to second home sites. One 1895 brochure was entitled "Vermont: Its fertile farms and summer homes" (Albers, 2000).

2 Byron and McClaughry (1989) advocate returning power to the towns and making Vermont a federation of "shires" – small localized units of government between a town and country in size.

3 The Quechee Lakes Corporation Development was also the first instance in Vermont where a developer was compelled to create and present a master open-space plan upon evidence that full buildout was contemplated and that environmental and community resources were committed. The plan was approved in 1986 and 1987, and is used in evaluating any new developments within the over 1000 acres of resort lands, ensuring that open space and natural areas are maintained.

4 The Environmental Board has long been interested in informal conflict resolution and began experimenting with dispute resolution and mediation in the late 1980s. On July 1, 2001, the board instituted a three-year mediation project available without charge to Act 250 participants.

5 The Environmental Board master permit policy contains the following footnote addressing the issue of 'vested rights' in reliance on partial findings used in master plan review: *Pursuant to Environmental Board and Supreme Court precedent and 10 V.S.A. Section 6086(b) relating to vested rights, it is generally not possible for a district commission to make final findings of fact and conclusions of law for a phased project under certain criteria, including criteria 5, 6, 7, 8, 9(A), 9(K), and 10, until a final decision is issued for a particular phase or for the entire project*

based upon the review of a complete application. Under current law, partial findings of fact can be appealed immediately or upon issuance of a final decision on a complete application. Rule 21 contemplates that partial findings of fact and conclusions of law are considered final for the purposes of appeal only and shall be binding for a reasonable period of time or until a final decision is issued based upon a complete application unless there is a "material" or "substantial change" or if the background facts have changed significantly. Parties may elect to reserve their appeal rights until final action on the complete application. Therefore, findings of fact and conclusions of law do not achieve true finality with vested rights until there has been final action on a complete application and all appeal issues have been fully litigated, if appeal rights are exercised. See 10 V.S.A., Section 6086(b), Rule 21(D), Rule 34 (B) and (C). Please note that Section 6086(b) was amended during the pendency of the Vermont Supreme Court decision, In re Taft Corners Associates, Inc., 160 Vt. 583 (1993), which held that unappealed decisions are final and cannot be attacked at a later date. The Court's analysis was based in part on the prior statutory language. For a good discussion of "completeness of application" which has a direct bearing on the "finality of decision" issue please refer to Vermont Supreme Court decisions: In re Agency of Administration, 141 Vt. 158 (1981) and In re Vermont Gas Systems, Inc., 150 Vt. 34 (1988), both of which have held that Act 250 jurisdiction does not formally attach until detailed plans indicate that 'construction is about to commence'. However, these decisions do not support the notion that the district commission or the Board is without authority to review applications for partial findings of fact pursuant to Rule 21(D) and 10 VSA Section 6086(b).

References

Albers, Jan (2000), *Hands on the Land: A History of the Vermont Landscape*, MIT Press, Cambridge, Massachusetts.

Bourdon, Don (1995), Director, Two Rivers Regional Planning Council, Woodstock, Vermont, personal interview (July 14).

Brooks, Richard, with Leonard, K. and Student Associates (1997), *Towards Community Sustainability: Vermont's Act 250, Volume II: The History, Plans, and Administration of Act 250*, Serena Press, South Royalton, Vermont

Byers, N. Gail and Wilson, Leonard (1983), *Managing Rural Growth: The Vermont Development Review Process*, State of Vermont, Environmental Board, Montpelier, Vermont.

Byron, Frank and McClaughry, John (1989), *The Vermont Papers: Recreating Democracy on a Human Scale*, Chelsea Green Publishing Company, Chelsea, Vermont.

Cannon, F., (1993), Economic Growth and the Environment. *Economic and Business Outlook*, Bank of America, NY, June, 1993.

Courtney, Elisabeth (1999), 'Taking Care', *Vermont Environmental Report*, Vermont Natural Resources Council, Montpelier, Vermont, Summer, 1999, p. 2.

Daniels, T. L. and Lapping, M. B. (1984), 'Has Vermont's land use control program failed?: evaluating Act 250', *American Planning Association Journal*, Autumn, pp. 502-508.

Defoe, Darren (1992), *Act 250: A Positive Economic Force for Vermont*, The Vermont Natural Resources Council, Montpelier, Vermont.

Dillon, John (1994), 'Balancing Act On A Tightrope: A Product of Boom Times, Act 250 Feels Heat In Recessions', *Rutland Herald*, Vol 19, No. 25, Sunday, April 3, 1994, pp. 1, 6.

Fulton, William, Pendall, Rolf, Nyguyen, Mai and Harrison, Alicia (2001), *Who Sprawls Most? How Growth Patterns Differ Across the U. S.*, Center on Urban and Metropolitan Policy, The Brookings Institution, Survey Series, Washington, DC, July 2001.

Governor's Commission on Vermont's Future (1988), *Report of the Governor's Commission on Vermont's Future: Guidelines for Growth*, State of Vermont, Montpelier, Vermont.

Hamilton, Susan (1986), *Corporate Land Speculation in Vermont: A Profile of Causes, Environmental Impacts, and Case Examples of the Rapid Subdivision of Rural Vermont Land*, Vermont Natural Resources Council, Fall 1986, Montpelier, Vermont.

Hamilton, Susan and Clark, Susan (1986), 'Parcellizing Vermont', *Vermont Environmental Report*, Vermont Natural Resources Council, Montpelier, Vermont, Fall, pp. 10-12.

Hensel, April (1996), District Coordinator, Environmental Board, North Springfield, Vermont, personal interview, September 4, 1996.

Klyza, Christopher McGrory and Trombulak, Stephen C. (1999), *The Story of Vermont: A Natural and Cultural History*, Middlebury College, Middlebury, Vermont.

Lake Champlain Basin Program (1996), *Opportunities for Action: An Evolving Plan for the Future of Lake Champlain*, LAKE3 Champlain Management Conference, Publication Series, Grand Isle, Vermont

Lapping, Mark (1978), 'The Nickle-and-Diming of Vermont Farmland', *Rutland Herald and Sunday Times*, April 4, 1978.

Lapping, Mark (1982), 'Landscape Planning in Vermont: An Overview', *Landscape Planning*, Vol. 8(4), pp. 349-362.

Lapping, Mark, Penfold, George E. and MacPherson, Susan, (1983), 'The Right-to-Farm Laws: Will They Reseolve Land Conflicts?', *Journal of Soil and Water Conservation*, Vol. 38(6), pp. 465-467.

Lindner, Will (1999), 'Sacred Cows: Vermont's Farms and Lake Come in Conflict', *Vermont Environmental Report*, Vermont Natural Resources Council, Montpelier, Vermont, Summer, 1999, pp. 15-20.

McClaughry, John (1975), 'The New Feudalism', National Conference on Federal Land Use Planning, Lewis and Clark College, Portland Oregon, March 21-22, 1975.

McCormack, Richard J. (1994), 'Senators Discuss Act 250 Process, Environmental Board: Primer', *The Sunday Rutland Herald and The Sunday Time Argus*, February 6, 1994, Section C, p. 3.

Merrill, L. S. (1987), The Road Not Taken, *Planning*, 53(11), pp. 22-24.

Olinger, David (1993), 'Vermont: A State Endangered', *Times*, Sunday, July 25, 1993, pp. 1, 6D.

Pfeiffer, Bryan and Dillon, John (1994), 'For Some, It's the Appeal: Rethinking Act 250', *Rutland Herald*, Vol. 138, No. 81, Monday, April 4, 1994, pp. 1, 10.

Sanford, R. M. and Stroud, H. B. (1997), 'Vermont's Act 250 Legislation: A Citizen-based Response to Rapid Growth and Development', *Land Use Policy*, 14(4), pp. 239-256.

Two Rivers-Ottauquechee Regional Commission, Southern Windsor Country Regional Planning Commission and Upper Valley Lake Sunapee Regional Commission (2001), *Upper Valley Housing Summit: Strengths, Weaknesses, Opportunity and Threats*, Two Rivers-Ottauquechee Regional Commission, Southern Windsor Country Regional Planning Commission and Upper Valley Lake Sunapee Regional Commission.

Vermont Act 249 Environmental Conservation Law, Title 18 Chapter 23; Title 10, Chapter 161. Montpelier, Vermont, (Amended, effective May 4, 1990).

Vermont Act 250, Land Use and Development Law, Title 10, Chapter 151, Montpelier, Vermont, 1970.

Vermont Agency of Commerce and Community Development (2000), *Strategic Plan*, Agency of Commerce and Community Development, Montpelier, Vermont.

Vermont Department of Housing and Community Affairs (1988), *Planning Manual for*

Vermont Municipalities, Department of Housing and Community Affairs, Agency of Commerce and Community Development, Montpelier, Vermont.
Vermont Department of Housing and Community Affairs (1997), *Consolidated Plan for Housing and Community Development Programs*, Department of Housing and Community Affairs, Agency of Commerce and Community Development, Montpelier, Vermont.
Vermont Department of Housing and Community Affairs (1999), *History of Planning in Vermont*, Department of Housing and Community Affairs, Agency of Commerce and Community Development, Montpelier, Vermont.
Vermont Environmental Board (1995), *Twenty-fifth Anniversary Report, 1970-1995*, Vermont Environmental Board, Montpelier, Vermont.
Vermont Environmental Board (1998), *Master Permit Policy*, [Amended March 29, 2000], Environmental Board, Montpelier, Vermont.
Vermont Forum on Sprawl (1999), *Vermont Forum on Sprawl Action Plan*, Burlington, Vermont.
Vermont Housing Council (1998), *Vermont Housing Needs Assessment*, Vermont Housing Council, Montpelier, Vermont.
Vermont Housing Council (1999), *Summary of the Vermont Housing Needs Assessment 1998-2003*, Vermont Housing Council. Montpelier, Vermont.
Vermont Natural Resources Council (1989a), 'The Origins of Act 250: A Talk with Former Governor Deane C. Davis', *Vermont Environmental Report*, Vermont Natural Resources Council, Montpelier, Vermont, Fall, pp. 17-19.
Vermont Natural Resources Council (1989b), '20 Years of Act 250: What's the Difference?', *Vermont Environmental Report*, Vermont Natural Resources Council, Montpelier, Vermont, Fall, p. 20.
Vermont Natural Resources Council (1999), '1999 Legislative Wrap-up', *Vermont Environmental Report*, Vermont Natural Resources Council, Montpelier, Vermont, Summer, pp. 22-24.
Vermont Planning Council (1968), *Vision and Choice: Vermont's Future*, State of Vermont, Montpelier, Vermont.
Vermont Supreme Court (1994), In re Frank A. Molgano, Jr., Supreme Court Docket #93-017, Environmental Board Case 8B0468-EB, 5 Vermont Law Week 314, Nov. 10, 1994.
Wilson, Leonard U. (1999), 'Land Use, Planning, and Environmental Protection', in Sherman, Michael (ed.), *Vermont State Government Since 1965*, University of Vermont, Burlington, statutes and court cases cited.

Chapter 3

The Pinelands

Robert J. Mason

Situated nearly in the center of the densely populated northeastern megalopolitan corridor is a million-acre tract of pine-oak forest traditionally known as the Pine Barrens or Pines, and more recently recognized as the Pinelands National Reserve (Figure 3.1). One-third of the land within the reserve is publicly owned; the other two-thirds is in private hands. Approximately 700,000 people live in the Pinelands, though parts of the central Pines have remarkably low population densities. The Pinelands are home to two major aquifers and a great many species of rare and endangered flora and fauna. In 1981, only four days before Ronald Reagan became President, Interior Secretary Cecil B. Andrus approved a comprehensive regional zoning and regulatory plan that would be implemented by local, county, state, and federal governments. The Pinelands became the country's first, and is still its only, 'national reserve'. How is the Pinelands region defined and constructed? How has the regional plan, and the region it is meant to protect, fared over the past two decades? What are the future prospects?

Regional Significance and History

Unlike most national parks in America, the New Jersey Pinelands is an area with towns, farms, and an extensive history of human occupance and resource exploitation. The region is prized especially for its vast stores of groundwater, but also for its unique ecology and history. It is known by various names, with the traditional appellation 'Pine Barrens' implying a forbidding place whose soils are not conducive to agricultural production. In fact, parts of the Pine Barrens are exquisitely suited for cranberry and blueberry production. Many of the region's residents, known as 'Pineys,' are more apt to recognize their own part of the Pinelands than to regard themselves as residents of the larger ecoregion (Berger and Sinton, 1985; Mason, 1992). When they do refer to the region, 'Pines' is the term that often is used. The official term, embraced by the U.S. and New Jersey governments, is 'Pinelands'. This term is meant to free the place of the stigma associated with 'barrens'. The officiousness of the term is reflected in the way that residents often refer to the regional planning agency, the Pinelands Commission, simply as 'the Pinelands'.

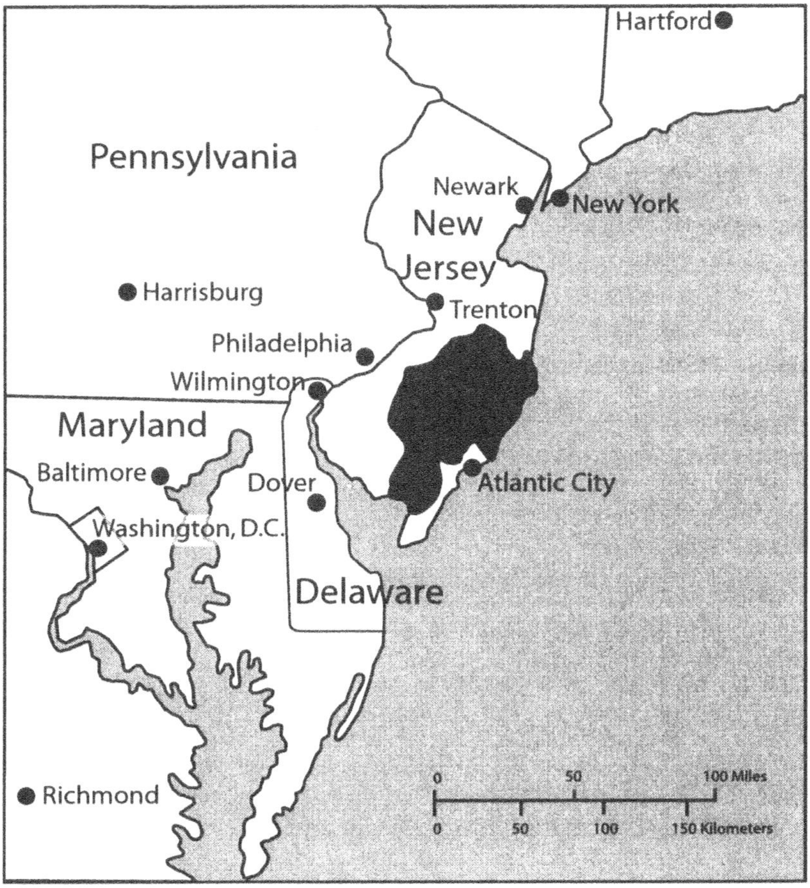

Figure 3.1 Pinelands Region, New Jersey

Water is central to the physical character and human history of the Pinelands, which are underlain by two major aquifers: the Kirkwood and the Cohansey. Collectively, these and several smaller aquifers contain about 17 trillion of gallons of pure water and supply 89 per cent of surface water flow. The entire Pinelands ecosystem, with its abundant wetlands, streams, and impoundments, is heavily dependent upon groundwater. Yet the aquifers can easily become contaminated, since they are overlain by sandy soils that are rather ineffective in filtering wastes (New Jersey Pinelands Commission, 1980; Rhodehamel, 1979; Trela and Douglas, 1979). Water quality is a prime concern for Pinelands protection advocates; so, too, is water quantity with particular attention to the potential for overuse and exploitation by outside interests. Indeed, in response to industrialist Joseph Wharton's plans to export water to Philadelphia, the New Jersey Legislature passed a law in 1884 that prohibited conveyance of water out-of-state. Even today, some residents harbor suspicions that the Pinelands plan is designed to preserve the groundwater resource for future large-scale industrial/commercial use.

Ecological and open-space protection also are driving concerns in Pinelands management. Indeed, some of the strongest advocates for regional protection have been ecologists (Forman, 1979; Sinton and Berger, 1985; Mason, 1992). The Pines are distinguished by extensive pine-oak forests, as well as cedar swamps and maple-gum-magnolia complexes. Of particular interest to naturalists are the pygmy pines, or Pine Plains, which drew the attention of conservationist Gifford Pinchot around the turn of the last century. These dwarf forests, occupying between 12,200 and 22,150 acres in total (McCormack, 1970) continue to engage ecologists, who debate the relative roles of fire frequency, soil conditions, and genetic factors in sustaining this unique forest ecosystem (McCormack, 1970; Good, et.al., 1979; New Jersey Pinelands Commission, 1980). The Pinelands also act as a sort of 'post-glacial arboretum' (Mason, 1992; p. 51), where a number of northern plants reach their extreme southern limits and southern plants reaches are at their northern extremes.

Pinelands cultural history and prehistory also are prominent factors in regional protection and planning, although not as central as hydrologic and ecological considerations. Earliest human occupance of the region has been established to be more than 10,000 years ago (Hartzog, 1982; New Jersey Pinelands Commission, 1980; Wacker, 1979; Sinton, 1982), while European settlement dates to the early 1700s (New Jersey Pinelands Commission, 1980). Over the past three centuries, the region has been intensively used. Through the mid-1800s, the Pinelands supplied wood for shipbuilding and for fueling urban furnaces and steam engines. Through the eighteenth and first half of the nineteenth centuries, the forests were cut to produce charcoal that stoked the iron industry. Limonite, or bog ore, was used to produce iron until the iron making industry moved westward toward the coal and concentrated iron ore deposits of the Midwest. Between 1800 and 1875, glassmaking thrived in the Pines. Other smaller industrial enterprises, including paper and textile mills, were present as well (New Jersey Pinelands Commission, 1986). In the early 1900s, Elizabeth White developed the cultivated blueberry at the town of Whitesbog, and during the course of the twentieth century, commercial cranberry and blueberry farming became significant agricultural industries. Until rather recently, the Pinelands landscape was a very settled and exploited one, with an extensive network of roads and small towns (Wacker, 1979). Many of these survive today only as ghost towns or as interesting place names such as Ong's Hat and Double Trouble that appear on maps but to the casual observer show no signs of human occupance.

The role of contemporary Pinelands culture is critical to our understanding of the region (Sinton, et al., 1979; Moonsammy, et al., 1987). 'Piney' is the term used to describe Pinelands natives, who are portrayed romantically by John McPhee (1967, 1974) and celebrated with 'Piney Power' signs and bumper stickers proudly displayed by Pinelands residents and non-residents alike. But despite the near-mythic status accorded Pineys, it is not at all clear who qualifies as one. According to Sinton (1981, p.40), a Piney is someone who 'is just a little deeper in the woods than you are'.

Pineys were not always viewed in flattering terms. Early in the 1900s, Elizabeth Kite (1913), based on her research at the Vineland Training School for Feeble Minded Boys and Girls, described Pineys as degenerate, lazy, and lustful. 'The real Piney,' wrote Kite, 'is a degenerate creature who has learned to provide for himself the bare necessities of life without entering into life's stimulating struggle...like the degenerate relative of the crab...kicking food into its mouth and enjoying the functioning's of reproduction, the Piney and all the rest of his type have become barnacles upon our civilization' (Kite 1913, p. 10). These unscientific impressions were reinforced by the ethnographic work of Kite's supervisor, Henry Goddard (1912). Although all of this in time would be debunked, the immediate effect was a call, in 1913, from Governor James F. Fielder for segregation and possible sterilization of this dangerous population (Goldstein, 1981).

If a Piney is regarded as someone who depends principally on the region's natural resources for his or her living, few true Pineys are to be found today. Yet Piney lore persists, and there is a rich tradition of music and story telling that is still commemorated. Bluegrass and old-time music can be heard weekly at Waretown's Albert Hall, home of the Pinelands Cultural Society. The Jersey Devil, a regional legend from the 1700s, is the namesake for today's hockey team based in northern New Jersey. Many of those who do regard themselves as Pineys, or who would be labeled as such by non-residents, are concerned about being regarded as 'museum pieces' (Mason, 1992). Moreover, although regional culture is intimately linked to the physical environment, representation and 'management' of local culture is not necessarily among the basic principles that guide ecosystem planning (see Yaffee, et al., 1996; Vogt, et al., 1997). Indeed, ecosystem management is a driving force in Pinelands management.

Given its complex pattern of human uses and physical characteristics, the Pinelands region may aptly be seen as many regions rather than one. Indeed, this is one of the continuing controversies in Pinelands regional planning: to what degree should the region be regarded as a whole region, versus a series of interlinked subregions (Berger and Sinton, 1985; Mason, 1992). The Pines have been mapped on the basis of hydrogeology, vegetation, and soils (Harshberger 1916; McCormick and Jones, 1973; Governor's Pinelands Review Committee, 1979; New Jersey Pinelands Commission, 1980). The formal planning process, described below, treats the Pinelands region essentially as a unified ecological region, delineated on the basis of a combination of physical factors. Arguably, though, the bedrock concern has been groundwater management, with other physical and cultural factors subsumed under the hydrologic umbrella.

Planning and Legislative History

As noted earlier, the State of New Jersey staked out a strong position on resource protection when in 1884 it passed a law prohibiting export of water out-of-state. This act was contemporary with New York State's establishment of the Adirondack Forest Preserve. New York's effort was driven by interest in protecting the Hudson River watershed to serve future industrial needs, though the recreation rationale was invoked as well (Graham, 1978; Liroff and Davis, 1981).

For many environmentalists, public ownership is the most certain way to provide for long-term protection of valued lands. For many landowners, acquisition is more desirable than regulation, while for elected officials it often is more politically palatable, but limited in utility by its costliness. The first major Pinelands public land designation was that of Lebanon State Forest, in the northern Pines, in 1908. A 1915 ballot question on acquisition of the Wharton tract, in the center of the Pines, was rejected by voters. Not until 1950 was the 100,000-acre Wharton National Forest established. Several other significant parcels also came into public ownership between 1905 and 1954, and more recently, the Pinelands Commission has carried out its own land acquisition program in conjunction with the New Jersey Department of Environmental Protection's Green Acres office and with additional funding from the federal government. Today, about one-third of the Pinelands and this includes federal military facilities, is in public ownership.

Although scientific interest in the Pinelands has been in evidence for more than a century, the first serious stirrings about regional planning did not come until about 1960, when the Pinelands Regional Planning Board came into being. Largely geared toward economic development in portions of Burlington and Ocean Counties, the Board's primary accomplishment was to commission planning studies (Herbert H. Smith Associates, 1963, 1964). The planning area was considerably smaller than today's Pinelands National Reserve.

More widespread concern about Pinelands protection was prompted by a 1967 proposal to build a huge jetport, to serve the New York region, in the Pinelands. This was a NIMBY (Not In My Backyard), or LULU (Locally Unwanted Land Use), of enormous proportion. Consideration of the Great Swamp site, located in northern New Jersey, had been brought to a halt by a powerful campaign mounted by the area's wealthy and influential citizens (Cavanaugh, 1978). The Pine Barrens jetport scheme, mimicking visionary planning schemes of decades earlier, called for a garden city surrounded by open space and connected with New York City and Philadelphia by high-speed rail. Needless to say, there was strong opposition from conservationists as well as the military, which had its own plans for the region (Goldstein, 1981). The issues took on statewide importance; indeed there was general opposition to a jetport being sited anywhere in New Jersey (Mason, 1992). In the end, the Pines proposal may have been done in as much by economic considerations as by political and environmental concerns.

The jetport era marked the beginning of more serious and widespread attention to Pinelands protection. It was around this time that John McPhee's (1967) popular book, *The Pine Barrens*, brought much attention to the region. A Department of the Interior study, commissioned in the late 1960s, was greeted locally with apprehension about federal involvement in the region and provoked a response in the form of the Pine Barrens Advisory Committee, a body appointed by the Burlington and Ocean County Boards of Chosen Freeholders. In 1969, a report released by the National Park Service (U.S. Department of the Interior, n.d.) set forth several alternative regional management schemes, which would ultimately be reflected in some of the debates over Pinelands management a decade later. The report described these management alternatives: 1) two national scientific reserves, one considerably larger than the other and with federal intervention in the form of land acquisition and regulation, 2) a state forest or national recreation area that relies on both conservation easements and land acquisition, and 3) a state Pinelands region that would rely on zoning and land acquisition. The last of these, which would involve a regional governing body representing towns, counties, and the federal and state governments, is the one that most closely resembles Pinelands management in its current form.

Although the Pine Barrens Advisory Committee endorsed the Park Service land acquisition proposals, the time was not ripe for a comprehensive top-down plan. Instead, a locally-based planning organization, the Pinelands Environmental Council (PEC) was authorized by state legislation. But it principally represented local and county-level interests. The PEC's jurisdiction was limited to the central Pinelands and it only had review and advisory powers; it could delay but not halt major new projects (Mason, 1992). It was similar in composition and powers to the much-criticized Tahoe Regional Planning Commission (Constantini and Hanf 1973; Strong 1984). The PEC did produce a *Plan for the Pinelands* (Pinelands Environmental Council, 1975), but its extensive reliance on half-acre zoning led New Jersey Commission of Environmental Protection David Bardin to characterize it as a 'developer's dream' (Goldstein 1981, p.106).

In contrast to emphasis on more concentrated development found in the sprawl management and later 'smart growth' literature (see Real Estate Research Corporation, 1974; Bank of America, 1995; Young, 1995; Leo, et al., 1998; Burchell, et al., 1999; Daniels, 1999; Zinn, 1999), the PEC recommended that:

> No owner of a parcel of land should have the privilege of utilizing his land so intensely that if all similar parcels were so utilized, the region would suffer degradation. Otherwise, a time may come when no further development may be permitted to occur, unfairly punishing those who did not develop their land early. (Pinelands Environmental Council, 1975, p. 8).

In the late 1960s and early 1970s, regional planning was coming on line in Vermont, Florida, Oregon, and Colorado (Popper, 1981; DeGrove, 1984; Mason, 1992). In New York State's Adirondacks, a comprehensive private-land use management plan that in many ways closely resembles the one later adopted in the Pinelands, took effect in 1973 (Booth, 1987; Graham, 1978; Liroff and Davis, 1981; Terrie, 1985; 1997; Mason, 1995). But despite the attention being focused on the Pinelands at the time, the political climate was not yet right for comprehensive regional planning to advance beyond the conceptual stage.

For a time, the PEC's efforts, minimally-effective as they were, helped keep state and federal regulators at bay. Moreover, the Pinelands had no clear and compelling single issue around which to rally (in the Adirondacks, for example, it was proposals for major second-home developments in the early 1970s). Nor was there a forceful single advocate for regional planning at that time. And although the Pinelands are a treasured ecological and recreational resource, it lacks the star quality mountains, glaciers, lakes, and rushing streams that are the stuff of Sierra Club posters and films.

The political situation would soon change dramatically. In 1973, Brendan T. Byrne was elected Governor of New Jersey and he became a powerful advocate for the Pinelands. Strongly influenced by his close friend, the writer John McPhee, Byrne wanted Pinelands protection to become the accomplishment for which he would be remembered. Because New Jersey's Governor has extraordinary powers of patronage and appointment, a skilled governor--and Byrne fit this description--can have tremendous influence over the state Legislature (Salmore and Salmore, 1998).

The PEC, hobbled by its weak plan and scandals involving various officials (Mason, 1992) lost its state share of funding in 1974. The PEC continued to function for a time and made a last-gasp effort to reinvent itself in 1978, but by that time the tide had turned toward setting up a more comprehensive system of centralized planning.

In 1977, in the face of growing threats from proposed senior citizen communities, increasing recreational demands, anticipated growth associated with approval of casino gambling in Atlantic city, and planned development of Outer Continental Shelf oil and gas, to be transported via pipeline through the Pinelands (Governor's Pinelands Review Committee, 1979), Governor Byrne issued Executive Order 56, creating the Pinelands Review Commission (PRC). The PRC's draft report, released in 1978, called for establishment of a Pinelands Planning and Management Commission. With substantial input from the Rutgers University Center for Coastal and Environmental Studies, the PRC established boundaries for the planning area (Merrill, et al., 1978). The planning scheme with its core 'preservation area' within a larger 'protection area', embodies that advocated by the UN Man and the Biosphere Program for the worldwide network of biosphere reserves (Figure 3.2). Biosphere reserves are meant to be landscapes in which human inhabitants are an integral part of the ecosystem (see Batisse, 1982; 1997; West and Brechin, 1991; Lucas, 1992; Wells and Brandon, 1992; Solecki, 1994). The reserves are designated by MAB, but managed by national or

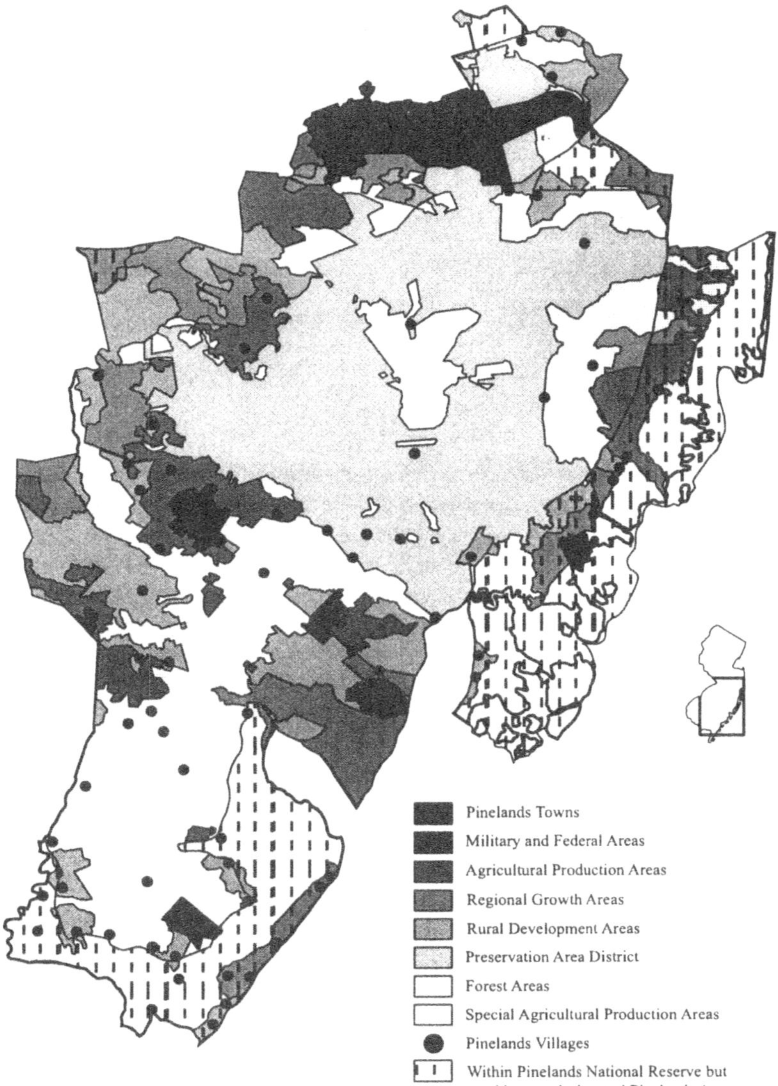

Figure 3.2 Pinelands Land Use Plan

sub-national governments. Ecological preservation is emphasized in the core area, while a carefully-defined range of human uses especially 'traditional use' is permitted in the buffer area. The buffer area generally is encircled by a transition zone, where more intense human uses co-exist with ecological protection. In 1982, the Pinelands gained international recognition when the Pinelands Biosphere Reserve was designated.

In the late 1970s, Pinelands protection was not just a New Jersey concern; the federal government was actively involved as well. The Mid-Atlantic Office of the National Park Service was involved with the PRC's efforts, while the federal Bureau of Outdoor Recreation as noted earlier produced a study that described several state-federal joint approaches to Pinelands management (US Department of the Interior, 1976). And there was considerable Congressional and Executive interest in promoting the 'greenline' approach to protected area planning. The greenline approach, in contrast with the more typical model of full government ownership and control, is one that favors 'public-private partnerships', with a mix of private and public landholdings. Conservation easements, regional zoning, and transfer of development rights are among the tools that might be employed. Private, local, state, and federal interests would be represented in the planning process (Hirner and Mertes, 1986; Belcher and Wellman, 1991). Although the Pinelands were to be the first formal greenline park in a system that would have been created under a bill proposed by New Jersey Senators Clifford Case and Harrison Williams, in the end the national greenline program never really did get off the ground. Still, several protected areas around the country embrace the general principles supported by greenline proponents (Little, 1983; Foresta, 1984; Mason, 1994). Chief among these areas are the Pinelands National Reserve and the Adirondack Park.

Two separate bills were written in the House: the Forsythe-Hughes bill, which favored local interests and encouraged municipalities to work cooperatively, and the Florio Bill, which gave greater weight to overriding regional interests. But neither bill was reported out of committee; instead, the Pinelands National Reserve was established as part of the lengthy National Parks and Recreation Act of 1978 (Van Abs, 1986). The legislation called for a 15-member planning body, with seven members appointed by the governor, seven members to come from the seven counties with land in the Pinelands, and a designee of the U.S. Secretary of the Interior.

Pinelands planning is enabled by state as well as federal law. New Jersey Executive Order 71, issued in 1979, established the 15-member planning body described above, and charged it with developing a comprehensive management plan for the million-acre area that had been defined by the Governor's Pinelands Review Committee (1979). In addition, Governor Byrne imposed a building moratorium that remained in effect until the Pinelands Protection Act was passed. A typical alignment of interest groups coalesced in response to the proposed legislation: farmers, local officials, and small entrepreneurs tended to oppose it, while environmentalists were in favor. Momentum, however, was with regional planning, and Byrne was able to sign a bill that called for completion of a comprehensive regional plan by August 8, 1980.

The plan was to be guided by the following general principles: For the inner Preservation Area:

1. Preserve an extensive and contiguous area of land in its natural state, thereby insuring the continuation of a Pinelands environment which contains the unique and significant ecological and other resources representative of the Pinelands area;
2. Promote compatible agricultural, horticultural, and recreational uses, including hunting, fishing, and trapping, within the framework of maintaining a Pinelands environment;
3. Prohibit any construction or development which is incompatible with the preservation of this unique area;
4. Provide a sufficient amount of undeveloped land to accommodate specific wilderness management practices, such as selective thinning which are necessary to maintain the special ecology of the Preservation Area; and
5. Protect and preserve the quantity and quality of existing surface and ground waters.

For the outer Protection Area:

1. Preserve and maintain the essential character of the existing Pinelands environment, including the plant and animal species indigenous thereto and the habitat therefore;
2. Protect and maintain the quality of surface and ground waters;
3. Promote the continuation and expansion of agricultural and horticultural uses.
4. Discourage piecemeal and scattered development.
5. Encourage appropriate patterns of compatible residential, commercial, and industrial development, in or adjacent to areas already utilized for such purposes, in order to accommodate regional growth influences in an orderly way while protecting the Pinelands environment from the individual and cumulative impacts thereof. (Mason, 1992, pp. 94-95)

The Comprehensive Management Plan

The Comprehensive Management Plan for the Pinelands (CMP) was developed quickly, under extraordinary pressure, and this fueled local resentment and the belief that extra-local interests were fully controlling the process. The New Jersey Assembly, concerned that the Governor's moratorium and the plan that was being developed were indeed too stringent, and too overbearing, strongly favored delay in approving the CMP. Intense negotiations between the Governor and legislators resulted in a compromise under which the plan for the Preservation Area was adopted as scheduled on August 8, 1980, while adoption of the Protection Area portion of the plan was delayed until November 14. Federal approval came in January 1981, when the outgoing Carter-administration Secretary of the Interior Cecil D. Andrus signed off on the CMP.

Central to Pinelands planning is the notion of a single Pinelands region, within which a predictable overall level of growth and development will occur. Areas of greatest environmental vulnerability will be afforded the strongest protections, while growth will be directed toward those parts of the region with the environmental capacity to accommodate it. The many cultural characteristics of the Pinelands are inventoried in the CMP, but given little planning attention beyond protections for sites of archaeological value and the so-called 'Piney exemption', described in the following paragraph.

The overarching basis for the CMP is protection of the Pinelands' critical ecological areas; this is operationalized in part through a stringent region-wide water quality standard for nitrate: two parts per million. The core Preservation Area is very strictly regulated, with emphasis on such uses as cranberry and blueberry cultivation, forestry, low-impact recreation, and limited resource extraction. Residential development is to be in and adjacent to designated towns, villages, and agricultural areas. For development outside these areas, applicants must demonstrate that they belong to a two-generation extended family with more than twenty years residence in the Pines or that they are dependent on 'employment or participation in a Pinelands resource-related activity' (New Jersey Pinelands Commission, 1980, p. 350); these activities are defined as 'including, but not limited to, forest products, berry agriculture and sand, gravel or minerals' (New Jersey Pinelands Commission, 1980, p.350). This is the 'Piney exemption'.

The Preservation Area, as well as the surrounding Protection Area, where development is allowed at varying densities, are subdivided into additional management areas.

They are as follows (Mason, 1992, pp. 99-100):

1. *Agricultural Production Areas*. Found in both the Preservation and Protection Areas, these zones consist of major areas devoted to agricultural uses, as well as adjacent lands so suited. Allowed uses are those related to agriculture, though municipalities have limited options for permitting other uses.
2. *Special Agricultural Production Areas*. Designated by municipalities in the preservation area, these are meant to protect areas devoted to berry production and native horticultural uses, as well as adjacent watershed lands.
3. *Military and Federal Installation Areas*. These are major existing federal landholdings.
4. *Forest Areas*. These are largely undeveloped areas that represent the 'essential character' of the Pinelands. Low-density residential development is permitted (average densities for each township are specified in the CMP), as well as certain other uses that would not greatly alter the character of these areas.
5. *Rural Development Areas*. These are meant to serve as buffers between more and less developed areas, as well as reserves for future development. Municipalities are afforded wide discretion in determining land uses in rural

development areas, though there is an overall density cap of 200 dwelling units per square mile. Within Rural Development Areas, municipalities have the option of designating 'Municipal Reserves'. Reserves are meant to absorb future growth beyond the capacity of existing regional growth areas (see below). Before an area's status can change from Reserve to Regional Growth, a series of environmental conditions must be met. Regional Growth Area standards then apply to the Reserve area.

6. *Regional Growth Areas*. These are in or adjacent to already developed areas, are experiencing growth pressures, and have been deemed capable of accommodating growth. The Commission allocates dwelling units and maximum densities for each town's Regional Growth districts. Regional Growth areas are meant to absorb growth demands generated by Atlantic City casino development, coastal growth pressures, and suburban expansion from the Philadelphia metropolitan area.
7. *Pinelands Towns and Villages*. These are existing settlements. Limited development is allowed in and around the center of the settlements, with a 3.2-acre maximum lot size for houses using conventional septic systems and a 1-acre minimum for those using alternative and innovative on-site treatment systems.

An additional 212,000 coastal acres are included in the Pinelands National Reserve and follow the same zoning scheme as the rest of the Pinelands National Reserve. But this area is under the jurisdiction of the New Jersey Division of Coastal Resources. While the Pinelands Commission provides comments on applications filed under CAFRA (Coastal Area Facilities Review Act), CAFRA regulations are, in fact, considerably less stringent than those that apply to the rest of the Pinelands.

The CMP includes a transfer of development rights (TDR) scheme. Sending areas are the Preservation Area, Agricultural Production Areas, and Special Agricultural Production Areas. Receiving areas are the Regional Growth Areas. Landowners in the Preservation Area receive one credit per 39 acres (prorated to the actual acreage on which rights are sold), except in wetlands areas, where the ratio is .2 credits per 39 acres and in Agricultural and Special Agricultural Production Areas, where it is .2 credits per 39 acres of upland, berry bogs, or fields and .2 credits per 39 acres of wetlands not being used for agriculture. Adoption of the TDR scheme was contentious. Supporters felt it provided necessary compensation and was a wise political move. Opponents argued that the compensation was unnecessary, since the CMP did not constitute a 'taking' of private property (see Randle, 1982).

Other key elements of the CMP include recommendations regarding wetlands, vegetation, forestry, agriculture, waste management, housing, and recreation. Critical to the plan's objectives is the region-wide implementation of the two-parts-per-million nitrate standard, aimed at protecting groundwater quality. On-site disposal systems anywhere in the Pinelands must be situated on lots of at least 3.2 acres, though exceptions can be made where alternative septic systems are used.

But after considerable experimentation with various systems, the Pinelands Commission has yet to find and endorse the ideal system for homeowners in this situation (New Jersey Pinelands Commission, 2000).

Plan Implementation

The Comprehensive Management Plan for the Pinelands (CMP) has been in place for two decades. It has adapted well to changing political circumstances and the Pinelands Commission has been able to give much greater attention to its public profile during its second decade than during its first several years.

In the early years, the Pinelands Commission was consumed with bringing about the conformance of local plans with the CMP and ensuring that the plan stood up to a variety of potentially destructive political and legal challenges. Conformance means that local master plans and zoning ordinances comply with the provisions of the CMP. While reasonable latitude is allowed in doing so, requirements regarding wetlands buffers, wastewater and septic standards, fire management provisions, and resource-extraction conditions are not negotiable. For those towns not in conformance, the Pinelands Commission is responsible for development review. For towns that have been certified, the Commission still can call up and review local decisions; it remains the ultimate decision-maker.

The number of conforming towns is viewed a key measure of the success of Pinelands planning. And, indeed, when compared with the Private Land Use and Development Plan for the Adirondacks, the Pinelands is a remarkable success story. All 53 Pinelands municipalities and seven counties are in conformance with the CMP, while in the Adirondack Park after nearly 30 years of plan implementation only 15 out of 103 municipalities have had their planning programs approved by the Adirondack Park Agency.

The Pinelands Commission made conformance a high priority and created a special 'conformance subcommittee' to work closely with individual municipalities in revising their plans and ordinances. The committee established strong working relationships with Pinelands towns; while underlying issues about management area boundaries, growth allocations, and local autonomy did not vanish, the convivial relationship between Commissioners and local representatives undoubtedly made the resolution of those issues more feasible and expedient than would have been the case with more formal, adversarial proceedings. Moreover, when a 1985 statewide bond issue was passed, $30 million became available for infrastructure projects in Pinelands Regional Growth Areas. Only towns that were in conformance with the CMP were eligible for the funds; furthermore, the county within which the town is situated also had to be in conformance. In 1987, a Pinelands Infrastructure Trust Fund was established to provide for wastewater treatment systems in Regional Growth to enable them to accommodate projected development.

Pinelands planning has also been made more palatable to local governments through provision of limited relief, through the Municipal Property Tax

Stabilization Act, in the form of payments to municipalities that contain large amounts of land on which development is prohibited. Payments in-lieu-of-taxes are a matter of course for towns in New York State's Adirondack Park, but this is not the case in the Pinelands. The relief provided in the Pinelands is short-term rather than ongoing relief. Furthermore, there has been contention about the fairness with which funds have been distributed (Mason, 1992). Still, the legislation was a step toward greater intra and inter (within New Jersey) regional equity.

Once the conformance process was largely completed and the concerns of most local governments reasonably satisfied, the Commission could turn its attention to a host of activities that had received rather scant attention up until this time. Public programs received renewed emphasis. The Commission expanded its outreach efforts and began to establish relationships with and publicize the activities of environmental and other organizations active in the region. In 1984, the Pinelands Education Advisory Council was established, as was the Pinelands Research and Management Council. Education programs, ecological research, and ongoing economic assessments of the CMP all were stepped up in the mid-1980s. These activities continue to consume a significant share of Pinelands Commission time and resources. In 2000, funding was approved for development of the Richard J. Sullivan Center for Environmental Policy and Education, which will act as a research and resource center for those with scholarly interests in the Pinelands.

The Commission continued to become more 'landowner and local government-friendly' over the years. Its early emphasis on towing the ecological line has gradually softened in the face of changing gubernatorial priorities and pressures from agricultural and development interests. This response is typical of maturing land-use programs elsewhere (see Popper, 1981; Mason, 1992). Many Pinelands localities now have 'Local Development Review Officers', whose job is to simplify the application process for those applying to the Pinelands Commission for permits to build single-family homes on existing lots. Steps also have been taken to simplify the application and permitting process for public entities. Recently, $23 million have been appropriated for state purchase of Pinelands Development Credits (PDCs), helping to boost the market price for the transferable credits. A separate program is in place to buy parcels of land that are too small to be developable. In 1997, a Rural Development Pilot Program was established. Its main objective is to examine economic development programs that might be applied in areas where growth potential is limited. Among the options under consideration are airport development, trade parks, and heritage tourism development.

Three major interests have been deeply involved in Pinelands planning and management: environmentalists, the development community, and agriculturalists. Environmental interests have long been represented through the New Jersey Conservation Foundation and local chapters of the Audubon Society. These and other local groups were dedicated and knowledgeable, but for many years yielded rather limited influence. During the early years of CMP development and implementation, major national groups, such as the Environmental Defense Fund,

Natural Resources Defense Council, and Sierra Club, were quite active. So too was the locally-based Friends of the Pine Barrens, which is now defunct. National interest has diminished, and most local and regional interests are now represented by the Pinelands Preservation Alliance. The Alliance is influential; its Board of Trustees includes former Governor Byrne (Honorary Chair), as well as former Pinelands Commission chair Franklin Parker. Also on the board are individuals representing key organizations and interests that have been active in Pinelands planning over the years. With a staff of eight, the PPA has become a powerful, influential voice in the planning process.

Another source of support for environmental interests is the Pinelands research community. Ecological concerns form the basis for much of the scientific research linked to Pinelands planning. The Commission has its own Science Office, which works closely with academic researchers at Rutgers, the State University of New Jersey, and other institutions. One of Commission's current major projects involves the development of environmental indicators for long-term environmental monitoring in the Mullica River watershed.

The development community was especially active during the period of plan development and implementation. The key actors were the Builders League of South Jersey (BLSJ) and the New Jersey Builders Association (NJBA). During the early years of Pinelands planning, the Coalition for the Sensible Preservation of the Pinelands was very active. Supported almost entirely by the BLSJ and NJBA, the Coalition lobbied the Legislature extensively, commissioned its own reports on Pinelands environmental issues, and even produced its own alternative plan for the Pinelands. Although its plan of course was not adopted, its participation did secure some concessions that favored development. The building community first through the Coalition, which has been inactive since the mid-1980s, and later through the BLSJ and NJBA was a key player in the politics of conformance and continues to be active in important development cases and when amendments to the CMP are taken up (Mason, 1992).

The New Jersey Builders Association currently is making the case against proposed reductions in density in some Regional Growth townships in the Pinelands. The Association contends that regional growth must be accommodated, and that although the CMP seeks to do so, state and local investments in infrastructure have not been sufficient to support this growth. NJBA is now imploring the Pinelands Commission to refrain from placing further restrictions or, alternatively, allow localities to place their own further restrictions on areas meant to accommodate future growth.

Agricultural interests have to some degree been allied with development interests. The agricultural community tends to be politically conservative and also well-represented politically in spite of its relatively small share in the state economy (Burch, 1975). Private property rights are a principal concern for the farm lobby in New Jersey, as is the case with farm lobbies elsewhere in the U.S. The New Jersey Farm Bureau has been particularly vigilant in Pinelands-related legislative lobbying and with the conformance process. Although the Pinelands Development Credit program sought to placate farmers, that was not the initial effect, if we are to judge by early reactions from the New Jersey Farm Bureau. Although some individual landholders have benefited rather handsomely (Mason, 1992), the overall benefits have been limited because until recently, the value of Pinelands Development Credits (PDCs) has been quite slow in appreciating. The Farm Bureau strongly supported allocation of farmland preservation funds from a dedicated state fund, as well as a new appraisal formula recently adopted by the Pinelands Commission, that seeks to increase the value of PDCs. Generally, though, the Farm Bureau has been a rather strong critic of Pinelands planning. However, the extent to which the Bureau represents Pinelands and state farmers as a whole is unclear; there are no survey data that speak to this question (Mason, 1992).

Interest group activity peaked during the period of plan development and early implementation in the late 1970s and early 1980s. Although these key interests continue to be active and well-represented, it seems that they have reached a level of accommodation with the Pinelands Commission and with each other that allows for a degree of remove from the day-to-day activities of Pinelands planning. Moreover, involvement has become more institutionalized and routine, as is typical of public participation in environmental affairs generally (Sewell and O'Riordan, 1976).

Pinelands planning has fared well through both Republic and Democratic administrations. The New Jersey Governor's office is the one institution most pivotal to the success of Pinelands planning; the Governor yields considerable power in appointing seven of the 15 Pinelands Commissioners and acting as the single most visible promoter for (or detractor to) the entire process. About three-quarters of the two decades of Pinelands planning has been under a Republican governor. When Republican Thomas Kean eked out a victory over environment-friendly James J. Florio in 1981, environmentalists feared for the worst. And indeed, Kean and later Republican Governor Christine Todd Whitman, were more accommodating of the development community's interests in the Pinelands than were Democrats Byrne and Florio. Whereas Governor Byrne's original seven appointees have been characterized as 'tree huggers', under Republicans Kean and Whitman, frequent calls from the development and agricultural communities for 'balance' and 'diversity' have been heeded. By this thinking, the ecological foundation for Pinelands management must be balanced with due consideration for economic development and administrative responsiveness to the needs of private landowners.

But New Jersey is a state where Republicans tend to be moderate and where environmental protection is a high priority issue (Mason and Mattson, 1990). While Ronald Reagan was attempting to dismantle the national environmental protection machinery in the early 1980s, Governor Kean upheld the protections in place in the Pinelands. It is unlikely that New Jersey voters would elect a governor, or put in place a legislative majority, that would be willing to undo Pinelands planning.

Yet Pinelands planning has been weakened or moderated, depending on one's perspective over the years. The program was, and to a considerable extent still is, managed by a dedicated staff of ecologists and environmental planners. The Pinelands Commission supports a team that is dedicated to scientific research and monitoring, and there is considerable research support from Rutgers University, the State University of New Jersey. The longstanding belief in ecological planning argues for close adherence to the environmental tenets of the CMP; compromise with local communities is at best a necessary evil. Of course, the complexion of the Commission itself determines how much compromise is ultimately made and clearly, the Commission has become less environmentally strident in recent years. This is evident in some of its responses to key current issues, described below.

Pinelands planning is now an accepted part of life in the region. There was a time when it was conceivable that it might go away. Many residents of Southern New Jersey saw the program as a land grab, a way for residents of the populous and politically influential northern part of the state to play without paying. Ultimately, the groundwater resources might be made available to outside industrial interests. During the late 1970s and early 1980s, there was considerable secessionist sentiment in southern New Jersey, and southern New Jersey's interests continue to play prominently in gubernatorial politics. While its population is small compared to that of the northern half of the state, it yields pivotal influence.

What limited evidence there is indicates that most Pinelands residents, as well as the great majority of New Jersey residents, support Pinelands planning. It is the vocal and effective voices of a relative few, acting on very deeply held convictions that may convey the impression that opposition is widespread. This impression is reinforced by wide media coverage of negative reaction to specific initiatives, such as a broadening of the Pinelands Commission's regulatory powers that was proposed in the early 1990s (Hajna, 1993). Disproportionate media coverage and accompanying political attention aside, locally based interests do raise very credible questions about the extent to which they are being asked to bear disproportionate burdens (development restrictions, diminished property values, loss of autonomy) in the service of the larger public interest.

Solecki's (1998) survey of Pinelands residents from a set of census tracts that includes both preservation and development-oriented zoning, along with areas that are experiencing growth as well as those whose populations are stable, indicated overall support for Pinelands planning of nearly 50 per cent, opposition of only 16 per cent, and a neutral or no opinion response from 35 per cent of the sample. Interestingly, even the strongest supporters tended to feel that the Pinelands plan has negative effects on the local and regional economy. This lends weight to the notion that support for regional planning is wide, if not very deep. Indeed, for many people it simply is not a primary concern. Yet the substantial group that holds no opinion about regional planning as well as those who are opinionated but perhaps only weakly so constitutes a crucial political bloc when it comes to voting for candidates in races where land use management is an issue, as well as in support for local and statewide referenda involving land protection.

Today, most interests recognize that Pinelands planning is here to stay and that they must accommodate themselves to it. Yet 2001 gubernatorial candidate Bret Schundler has proposed to do away with Pinelands planning. Schundler is a maverick Republican archconservative that ran against the state's Republican establishment and succeeded in winning the primary. Although it is unlikely he will be elected governor, he is a force in keeping alive the long-standing concerns of those southern New Jersey residents who view Pinelands planning as nothing more than a massive land grab.

Notable Recent Issues

One recent Pinelands issue that has attracted a good deal of attention is deeply rooted in the history of Pinelands planning. It involves J. Garfield DeMarco, a member of an Atlantic County family that has had a central role in local politics for decades (Mason, 1992). DeMarco is a major landowner and cranberry grower, active and influential in the Atlantic County Republican Party, and the first chairman of the Pinelands Environmental Commission during the 1970s. His brother Mark, as municipal attorney for Woodland Township, was engaged in protracted court battles involving the legality of Pinelands land transactions. Garfield's political activities have put him in good stead with current Department of Environmental Protection Commissioner Robert Shinn. This set the stage for recent controversy surrounding DeMarco's expansion of his cranberry lands. DeMarco's conversion of 22 acres of wetlands to cranberry bogs was done without the necessary permits, a violation of New Jersey's Freshwater Wetlands Protection Act. But DeMarco initially was not fined; instead a deal was negotiated that requires him to accept deed restrictions on 591 acres of forested land that he owns, in addition to donating 75 acres for inclusion in New Jersey's state park system. Although DeMarco was ultimately assessed a fine of about $600,000, he stands to realize an enormous benefit on the order of about $3 million from the sale of development credits on 3,600 acres of land in Woodland Township (Regan, 2001).

The DeMarco's have been frequent and vocal critics of Pinelands planning. Yet they are among its greatest beneficiaries. The Pinelands Development Credit program, in particular, has provided them with a windfall. Clearly, one can oppose a regulatory system and then reap as much benefit as possible from it. In this case, the extent to which their opposition has been sincere, as opposed to simply a matter of political posturing, is not entirely clear. But it would appear that in Woodland Township, where Garfield DeMarco has been most engaged politically, he as a local elite, has been an exceptional gatekeeper, very successfully shaping larger processes (i.e., regional growth and Pinelands planning) in ways that provided considerable local and personal benefit (Mason, 1992).

Another Pinelands issue of relatively recent vintage is that of cell tower placement. This, of course, was not anticipated when the CMP was developed. A 35-foot height restriction on structures applies through much of the Pinelands, with towns and Regional Growth Areas exempted. Under regulations adopted in 1995, the Commission can grant exemptions to this rule for cell towers. In those cases, the towers must not produce 'substantially detrimental' impacts, and must minimize visual impacts, be located in already disturbed areas, and be under 200 feet in height (N.J.A.C. 7:50-5.4 et seq.). In addition, the structures must be in conformance with a regional comprehensive plan for placement of cell towers. Such a plan was adopted by the Pinelands Commission in 1998, but had to be modified to take into account the ascendance of PCS technology. A chief objective of the plan is to meet regional telecommunications needs while erecting the smallest possible number of towers. The current plan, adopted early in 2000, allows for six new towers and use of 30 existing or previously-approved structures. The Pinelands Preservation Alliance vigorously opposed both plans, arguing that too many towers are permitted, public concerns have been ignored at the expense of industry interests, and structures are being permitted in ecologically-sensitive areas, such as the West Plains, where dwarf pines are present.

Pinelands Planning as Smart Growth

Much of the contention about Pinelands planning has involved limits that the plan places on growth. But since the plan's inception, there also has been concern about the burden of too much growth in those parts of the region to which the plan directs growth. Recent increases in the value and use of development credits in Regional Growth Areas, in conjunction with general concerns about the impacts of suburban growth have served to strengthen local sentiment favoring growth limitations. Indeed, Pinelands planning may be held up as a model for 'smart growth' (Real Estate Research Corporation, 1974; Bank of America, 1995; Young, 1995; Leo, et al., 1998; Burchell, et al., 1999; Daniels, 1999; Zinn, 1999), in that it accommodates, rather than resists growth, and directs that growth to areas deemed most suited for development. It stresses a cooperative, intergovernmental approach to regional planning just the sort of approach embraced by advocates of smart growth. But while Pinelands planning and smart growth efforts share certain fundamental precepts, there also is a major difference. Pinelands planning has behind it much more regulatory authority than does the typical smart growth effort; in no small measure, the threat of stringent enforcement of regulations can act as an inducement for cooperation among the parties involved.

The recent smart growth movement, which has taken hold especially in some of the major eastern metropolitan regions, not surprisingly has come home to roost in rapidly-growing parts of the Pinelands. Even though this growth is by CMP design, it is not entirely welcome. Several Pinelands townships that contain major Regional Growth Areas have approached the Pinelands Commission, asking for reductions in their growth allocations. Early on in the course of Pinelands plan implementation, many, though not all of these towns were willing to accept growth, and the Pinelands Commission faced its greatest difficulties in dealing with those towns who felt that they were being denied the opportunity to grow. Egg Harbor Township, in the eastern Pinelands, now argues that it has grown too much. Almost all of its land within the Pinelands boundary is zoned for growth at an average density of 3.5 units per acre, meaning a total of 33,725 houses in the Regional Growth part of the town. Local officials argue that development has been rapid, often of poor quality, and has burdened the town with undue fiscal and environmental impacts.

On the Pinelands Commission's calendar for late 2001 is an amendment to the CMP that would allow Egg Harbor and other Regional Growth townships greater flexibility in meeting these requirements, as well as reduction in the densities assigned to developable portions of their Regional Growth Areas. Part of this approach would be a 'timed growth' element that would control the pace of growth that these towns are obligated to accept. These measures have the qualified support of the Pinelands Preservation Alliance, which, as noted earlier, is the lead non-governmental organization that monitors and seeks to influence the Pinelands planning process.

Other recent Pinelands Commission actions, emblematic of a general shift on its part toward being more accommodating of development interests, have riled environmentalists. In 1999, the Commission approved an amendment to the CMP to allow redesignation of land zoned for agriculture under the CMP, so that a school can be constructed. The amendment, tailored to this specific situation and presented as a pilot program" allows for development of public educational facilities in Agricultural Production Areas and Rural Development Areas if certain conditions are met. Supporters of the rezoning contend that the area is currently a sod farm and not an important ecological resource, while opponents see this change as a precedent-setting weakening of the CMP. Former Secretary of the Interior Bruce Babbit granted the needed federal approval for the change, but with an admonition to the Commission to be cautious about sending up any more such amendments. Terrence Moore, who had been the Commission's Executive Director since its inception, did not approve of the zoning change and it is widely speculated that this is what prompted his resignation in 2000.

Environmentalists also have been outraged by the Commission's response to discovery of timber rattlesnakes in a Burlington County development named, interestingly enough, The Sanctuary. In 1995, when development was first approved, the endangered snakes had not been detected, though it had long been known that they are present in that part of the Pines. Construction was halted in 1998, when snakes were tracked on the site. The development is already partially completed, and under a settlement reached in 2000, the developer would be permitted to build an additional 147 homes. Snake dens would be fenced off and culverts constructed to allow snakes to travel under roads. The developer would also sell to the state 1,200 acres, in other parts of the Pinelands, for a sum exceeding $5 million. The Pinelands Preservation Alliance, along with the New Jersey Audubon Society and Natural Resources Defense Council, argues that the CMP mandates protection for endangered species, pure and simple. In this case, it seems, the Commission is unwilling to halt development because it fears it would lose an inevitable court battle. Moreover, snakes are not an easy species to defend. This leads to questions about whether or not residents of The Sanctuary would actually learn to co-exist with the snakes, as envisioned under the settlement that was reached.

Conclusion

Pinelands planning is an important model for guiding growth and protecting the integrity of ecosystems (Collins and Russell, 1988; Lilieholm and Romm, 1992). It appears that the CMP has been largely successful with respect to sustaining a regional landscape that fits the U.N. biosphere reserve model, with its core, buffer, and transition zones. Indeed, a recent analysis by Walker and Solecki (1999, p. 231) reveals that:

> The area with least conversion is the preservation core (in relative terms), and the other two Pinelands regions show decidedly less conversion than the outlying census tracts.

A less rigorous assessment, by Peterson (1999), contends that the Pinelands has neither boomed nor busted. It has grown steadily, prospered and stayed attractive. Still, concern has been growing as already noted about the pace of growth in designated growth areas. Furthermore, alarms are being raised about rapid growth in some areas just outside the Pinelands boundary (Mansnerus, 1998). Moreover, the value of Pinelands Development Credits, which transfer development rights from the core Preservation Area to designated Regional Growth Areas, has been increasing in recent times in response to increased pressures for development and a $23 million purchase program supported by funds from a statewide land preservation bond issue approved by voters in 1998. Collectively, these trends raise questions, as well as providing new opportunities, regarding the ability of the CMP to maintain the integrity of the Protection Area, as well as allow for a suitable transition zone from less-developed to more-developed parts of the Pinelands region.

Despite its overall success in managing Pinelands growth, the concerns just noted notwithstanding, Pinelands planning has trended in a more conservative direction, particularly over the course of the past decade. Pinelands Commission staff are more accommodating of development interest, i.e., a balance among competing interests and the ecocentric voice of Commission staff is not as influential as in prior years. The leading environmental organization, the Pinelands Preservation Alliance, remains an ally of the Commission but is alarmed at the increasing calls for balance. In its view, there is only one mandate: to protect, preserve, and enhance the Pinelands. That concern notwithstanding, the ecological basis for Pinelands planning is well-established and in all likelihood will not be eroded too deeply. Although the development and agricultural communities are very well-represented, there is no representative equal to the Pinelands Preservation Alliance for local interests who would want to rein in Pinelands planning. The Pinelands Municipal Council, long dormant, was revived in 1995, and it is the main voice for local communities. Its activities are limited principally to participation in periodic reviews of the CMP, and it appears that this will remain the case for the foreseeable future.

Pinelands management is an evolving process of conflict and accommodation among various interests, key among them agriculture, homebuilding, and environmental protection. In the end, the planning process did not give as much weight to local and subregional concerns as would be supported by those cultural geographers, historians, human ecologists, and others who are deeply concerned with the human character of the region (Rubenstein, 1983; Berger and Sinton, 1985; Mason, 1992). Even these supporters of local interests readily acknowledge, however, that achieving an appropriate balance between natural and cultural concerns is a very elusive prospect indeed. While many local residents want to protect the natural character of the Pinelands, they are wary of government intervention. Yet without strong planning controls, the region is extremely susceptible to the designs of well-funded developers who are represented by high-priced legal talent. Local planning boards can be so easily overburdened and overwhelmed by external interests, who, in many cases will work in conjunction with local interests who stand to benefit. Thus, external government intervention is needed if the region's natural and cultural integrity are to be protected. Pinelands planning has been quite successful with the former, less so with the latter. What remains to be seen is how the region will fare as the Pinelands Commission embarks on its next scheduled 5-year review of the CMP, as new development pressures emerge, and as political forces in New Jersey continue to realign.

References

Bank of America (1995), *Beyond Sprawl*, www.bankofmerica.com/community/comm._env_urban1.html.

Berger, J. and Sinton, J.W. (1985), *Water, Earth, and Fire: Land Use and Environmental Planning in the New Jersey Pine Barrens*, Johns Hopkins University Press, Baltimore.

Booth, R. (1987), 'New York's Adirondack Park Agency', in D. J. Brower and D. S. Carol (eds.), *Managing Land-Use Conflicts: Case Studies in Special Area Managementt*, Duke University Press, Durham, pp. 140-184.

Burch, P.H. (1975), 'Interest Groups', in R. Lehne and A. Rosenthal (eds.), *Politics in New Jersey*, Rutgers University Eagleton Institute of Politics, New Brunswick.

Burchell, R. W., Naveed, A. S., Listokin, D., Phillips, H., Downs, A., Seskin, S., Davis, J. S., Moore, T., Helton, D., and Fall, M. (1998), *The Costs of Sprawl Revisited*, National Academy Press, Washington, DC.

Cavanaugh, C. (1978), *Saving the Great Swamp: The People, the Power Brokers, and an Urban Wilderness*, Columbia Publishing, Frenchtown, NJ.

Collins, B.R. and Russell, E.W.B. (1988), *Protecting the New Jersey Pinelands: A New Direction in Land-Use Management*, Rutgers University Press, New Brunswick.

Constantini, E. and Hanf. K. (1973), *The Environmental Impulse and Its Competitors: Attitudes, Interests, and Institutions at Lake Tahoe*, University of California at Davis Institute of Governmental Affairs, Davis.

Daniels, T.A. (1999), *When City and Country Collide: Managing Growth in the Metropolitan Fringe*, Island Press, Washington, DC.

DeGrove, J.M. (1984), *Land, Growth & Politics*, American Planning Association, Chicago.

Foresta, Ronald A. (1984), *America's National Parks and Their Keepers*, Resources for the Future, Washington, DC.

Forman, R.T.T., ed. (1979), *Pine Barrens: Ecosystem and Landscape*, Academic Press, New York.

Goddard, H.H. (1912), *The Kallikak Family*, Macmillan, New York.

Goldstein, J. (1981), *Environmental Decision Making in Rural Locales: The Pine Barrens*, Praeger, New York.

Good, R.E., Good, N.F., and Anderson, J.W. (1979), 'The Pine Barren Plans', in R. T. T. Forman (ed.), *Pine Barrens: Ecosystem and Landscape*, Academic Press, New York, pp. 283 -295.

Governor's Pinelands Review Committee (1979), *Planning and Management of the New Jersey Pinelands*, State of New Jersey, Trenton.

Graham, F.J., Jr. (1978), *The Adirondack Park: A Political History*, Knopf, New York.

Hajna, L.R. (1993), 'Pinelands Bill Triggers Debate on Panel's Future', *Courier-Post*, 18 October, 1A, 4A.

Harshberger, J. W. (1916), *The Vegetation of the New Jersey Pine Barrens: An Ecologic Investigation*, Christopher Sower, Philadelphia.

Hartzog, S. (1982), 'Pine Barren's Prehistory: Current Directions', in J. W. Sinton (ed.), *History, Culture, and Archeology of the Pine Barrens: Essays from the Third Pine Barrens Research Conference*, Stockton State College Center for Environmental Research, Pomona, N J, pp. 3-5.

Herbert H. Smith Associates (1963), *The New Jersey Pinelands Region*, Herbert H. Smith Associates, West Trenton, NJ.

Herbert H. Smith Associates (1964), *Future Development Plans: The New Jersey Pine Barrens Region*, Herbert H. Smith Associates, West Trenton, NJ.

Hirner, D.K. and Mertes, J.D. (1986), 'Greenlining for Landscape Preservation', *Parks and Recreation*, Vol. 21, pp. 30-34, 59.

Kite, E.S. (1913), 'The Pineys', *Survey* Vol. 31, pp. 7-13, 38-40.

Leo, C., Beavis, M.A., Carver, A., and Turner, R. (1998), 'Is Urban Sprawl Back on the Political Agenda? Local Growth Control, Regional Growth Management, and Politics', *Urban Affairs Review*, Vol. 34, pp. 179-212.

Lilieholm, R.J. and Romm, J. (1992), 'Pinelands National Reserve: An Intergovernmental Approach to Nature Preservation', *Environmental Management*, Vol. 16, pp. 335-343.

Liroff, R.A., and Davis, G.G. (1981), *Protecting Open Space: Land Use Controls in the Adirondack Park*, Ballinger, Cambridge, MA.

Little, C,E. (1983), 'The National Perspective: Greenline Parks', in New York State Department of Environmental Conservation, *Proceedings: Greenline and Urbanline Parks Conference*, New York State Department of Environmental Conservation, Albany, NY, pp. 3-5.

Lucas, P.H.C. (1992), *Protected Landscapes: A Guide for Policy-makers and Planners*, Chapman and Hall, London.

McCormick, J. (1970), *The Pine Barrens: A Preliminary Ecological Inventory*, New Jersey State Museum, Trenton.

McCormick, J. and Jones, L. (1973), *The Pine Barrens: Vegetation Geography*, New Jersey State Museum, Trenton.

McPhee, J. (1967), *The Pine Barrens*, Farrar, Straus, and Giroux, New York.

McPhee, J. (1974), 'People of the New Jersey Pine Barrens', *National Geographic*, Vol. 45, pp. 52-77.

Mansnerus, L. (1998), 'Pine Barrens on Edge', *New York Times–New Jersey*, 3 May, 1, 12.

Mason, R.J. (1992), *Contested Lands: Conflict and Compromise in New Jersey's Pine Barrens*, Temple University Press, Philadelphia.

Mason, R.J. (1994), 'The Greenlining of America', *Land Use Policy*, Vol. 11, pp. 208-221.

Mason, R.J. (1995), 'Sustainability, Regional Planning and the Future of New York's Adirondack Park', *Progress in Rural Policy and Planning*, Vol. 5, pp. 15-28.

Mason, R.J. and Mattson, M.T. (1990), *Atlas of United States Environmental Issues*, Macmillan, New York.

Merrill, L.G., Jr., et al. (1978), *A Plan for Pinelands National Preserve*, Rutgers University Center for Coastal and Environmental Studies, New Brunswick, NJ.

Moonsammy, R.Z., Cohen, D.S., and Williams, L.E., eds. (1987), *Pinelands Folklife*, Rutgers University Press, New Brunswick.

New Jersey Pinelands Commission (1980), *Comprehensive Management Plan for the Pinelands National Reserve and Pinelands Area*, New Jersey Pinelands Commission, New Lisbon, NJ.

New Jersey Pinelands Commission (1986), *Pinelands Cultural Resource Management Plan for the Historic Period Sites*, New Jersey Pinelands Commission, New Lisbon, N J.

New Jersey Pinelands Commission (2000), 'Ad Hoc Committee On Alternative Septic Systems Staff Report', 28 April.

Peterson, J.E. (1999), 'Smart Growth's Fiscal Lab', *Governing Magazine*, Vol. 12, pp. 36-42.

Pinelands Environmental Council (1975), *Plan for the Pinelands*, Pinelands Environmental Council, Browns Mill, N J.

Popper, F.J. (1981), *The Politics of Land-Use Reform*, University of Wisconsin Press, Madison.

Randle, E. (1982), 'The National Reserve System and Transferable Development Rights: Is the New Jersey Pinelands Plan an Unconstitutional 'Taking'?', *Boston College Environmental Affairs Law Review*, Vol. 10, pp. 183-241.

Real Estate Research Corporation (1974), 'The Costs of Sprawl: Environmental and Economic Costs of Alternative Residential Development Patterns at the Urban Fringe', Council on Environmental Quality, Washington, DC.

Regan, T.L. (2000), 'State Pays GOP Insider $3M for Pinelands Acres', *Trenton Times*, 27 July, 3.

Rhodehamel, E.C. (1979), 'Hydrology of the New Jersey Pine Barrens', in R. T. T. Forman (ed.), *Pine Barrens: Ecosystem and Landscape*, Academic Press, New York, pp. 147-167.

Rubinstein, N.J. (1983), *A Psycho-Social Impact Analysis of Environmental Change in New Jersey's Pine Barrens*, Ph.D. dissertation, University Microfilms, Ann Arbor, MI.

Salmore, B.G. and Salmore, S.A. (1998), *New Jersey Politics and Government: Suburban Politics Comes of Age*, 2nd ed., University of Nebraska Press, Lincoln and London.

Sewell, W.R.D. and O'Riordan, T. (1976), 'The Culture of Participation in Environmental Decision-Making', *Natural Resources Journal*, Vol. 19, pp. 337-358.

Sinton, J.W., ed. (1979). *Natural and Cultural Resources of the New Jersey Pine Barrens: Inputs and Research Needs for Planning. Proceedings and Papers of the First Research Conference on the New Jersey Pine Barrens*, Stockton State College Center for Environmental Research, Pomona, NJ.

Sinton, J.W. (1981), 'Cultural Self-Preservation: Planning for Local Cultures in the New Jersey Pine Barrens', *New Jersey Folklore*, Vol. 2, pp. 39-42.

Solecki, W.D. (1994), 'Putting the Biosphere Reserve Concept into Practice: Some Evidence of Impacts in Rural Communities in the United States', *Environmental Conservation*, Vol. 21, pp. 242-247.

Solecki, W.D. (1998), 'Local Attitudes on Regional Ecosystem Management: A Study of New Jersey Pinelands Residents', *Society and Natural Resources*, Vol. 11, pp. 461-463.

Strong, D.H. (1984), *Tahoe: An Environmental History*, University of Nebraska Press, Lincoln and London.

Terrie, P. (1981), *Forever Wild: Environmental Aesthetics and the Adirondack Forest Preserve*, Temple University Press, Philadelphia.

Terrie, P. (1997), *Contested Terrain: A New History of Nature and People in the Adirondack*, Syracuse University Press, Syracuse.

Trela, J. and Douglas, L. (1979), 'Soils, Septic Systems and Carrying Capacity in the Pine Barrens', in J.W. Sinton (ed.), *Natural and Cultural Resources of the New Jersey Pine Barrens: Inputs and Research Needs for Planning, Proceedings and Papers of the First Research Conference on the New Jersey Pine Barrens*, Stockton State College Center for Environmental Research, Pomona, NJ, pp. 37 - 58.

U.S. Department of the Interior, National Park Service (1976), *New Jersey Pine Barrens: Concepts for Preservation*, Mimeographed.

U.S. Department of the Interior, National Park Service. (n.d.), *Pine Barrens of New Jersey*, Study Report, Mimeographed.

Van Abs, Daniel J. (1986), *Regional Environmental Management in the Pinelands National Reserve*, Ph.D. dissertation, University Microfilms, Ann Arbor, MI.

Vogt, K.A., Gordon, J.C., Wargo, J.P., Vogt, D.J., Asbjornsen, H., Palmiotto, P.A., Clark, H.J., O'Hara, J.L., Keeton, W.S., Patel-Weynard, T., and Witten, E. (1997), *Ecosystems: Balancing Science with Management*, Springer-Verlag, New York.

Wacker, P.O. (1979), 'Human Exploitation of the New Jersey Pine Barrens Before 1900', in R.T.T. Forman (ed.), *Pine Barrens: Ecosystem and Landscape*, Academic Press, New York, pp. 3 - 23.

Walker, R.T. and Solecki, W.D. (1999), 'Managing Land Use and Land-Cover Change: The New Jersey Pinelands Biosphere Reserve', *Annals of the Association of American Geographers*, Vol. 89(2), pp. 220-237.

Wells, M. and Brandon, K. (1992), *People and Parks: Linking Protected Area Management with Local Communities*, The World Bank, The World Wildlife Fund, and US Agency for International Development, Washington, DC.

West, P.C. and Brechin, S.R., eds. (1991), *Resident Peoples and National Parks: Social Dilemmas and Strategies in International Conservation*, University of Arizona Press, Tucson, AZ.

Yaffee, S.L., Phillips, A.F., Frentz, I.C., Hardy, P.W., Maleki, S.M. and Thorpe, B.E. (1996), *Ecosystem Management in the United States*, Island Press, Washington, DC.

Young, Dwight (1995), Alternative to Sprawl, Lincoln Institute of Land Policy, Cambridge, MA.

Zinn, J. A. (1999), Conserving Land Resources: The Clinton Administration Initiatives and Legislative Action, Congressional Research Service, Washington, DC.

Chapter 4

The Niagara Fruit Belt: Planning Conflicts in the Preservation of a National Resource

Hugh J. Gayler

The Niagara Fruit Belt stands as an important symbol to the long and difficult struggle to preserve a unique agricultural resource in Canada. This area in the northern part of the Niagara Region is relatively small, extending some 25 miles along the southern shore of Lake Ontario between Grimsby and the Niagara River, and extending inland between one and seven miles from the lakeshore to the Niagara Escarpment (Figure 4.1). However, with its important combination of soil and microclimatic conditions, it is one of only three small areas in Canada (the other two being the Okanagan Valley in British Columbia and parts of Essex County in south-western Ontario) that is capable of supporting a large-scale, commercial, tender-fruit and grape growing industry; and while this particular agricultural industry is relatively unimportant on the Canadian scene, let alone the wider North American or World scene, it has come to represent a source of pride in Ontario in the face of considerable hurdles. From being a backwater producing fruits for the fresh and processing markets and making notoriously bad wines, the Niagara Fruit Belt has emerged in the last twenty years as a quality wine-producing region, spawning in turn a growing agrotourism industry. Meanwhile, the tender-fruit industry has adjusted to the competition from the United States under the North American Free Trade Association (NAFTA) and is itself now in a growth phase; and a third industry, horticulture, competing for much the same land area, has seen a doubling of the area under glass in the last ten years, resulting in further efforts to preserve this agricultural resource.

The long and difficult struggle regarding the preservation of some of Canada's best agricultural land results from the Niagara Fruit Belt lying at the western end of the Golden Horseshoe, the name given to an extensive area of urban development surrounding the Greater Toronto Area (GTA) in southern Ontario and constituting one of Canada's fastest growing urban areas with a population of nearly six million people. While the Niagara Region today may have one of the lowest growth rates within this Golden Horseshoe area, it has a political and planning history that does not inspire confidence that this unique agricultural resource will be adequately protected. Since the Second World War, the Fruit Belt, like so many areas of

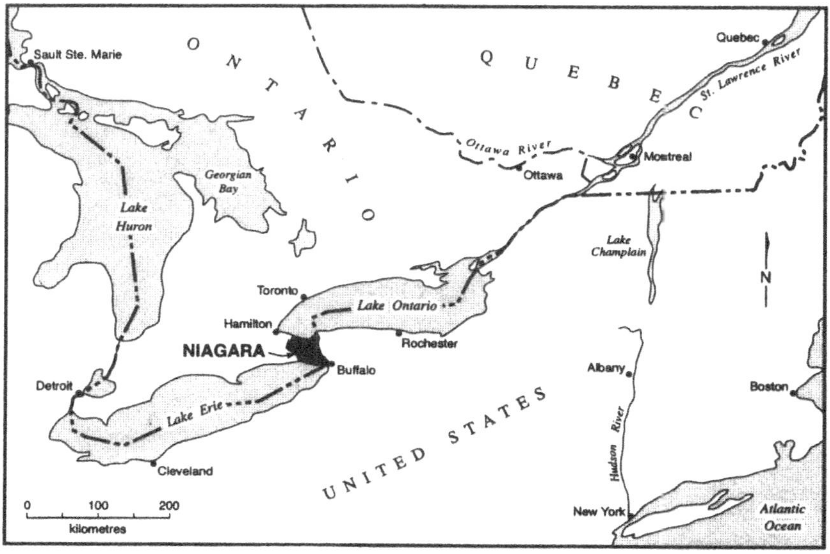

Figure 4.1 Location of the Niagara Region

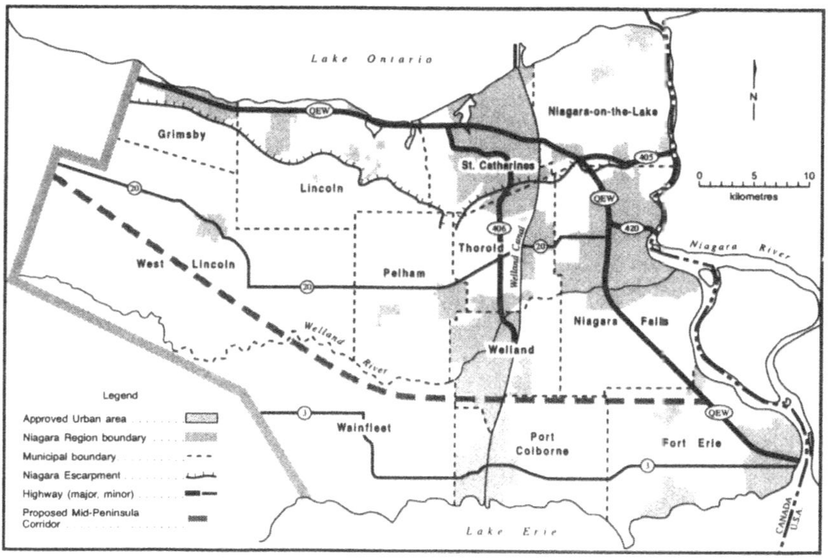

Figure 4.2 Niagara Urban Areas, 2002

North America, has been subjected to low-density, car-dependent suburban development about its two major industrial cities, Hamilton and St. Catharines, as well as some of the smaller communities such as Grimsby and Niagara-on-the-Lake. This suburban expansion in North America was not simply fuelled by a boom in population growth in the 1950s and 1960s, but also a desire on the part of millions of aspiring middle-class families to buy into a utopia that promulgated home-ownership, and more especially the detached, single-family house on a spacious lot. The suburb invoked images of country living, in a setting that was new, progressive, healthy and good for growing families and consequently divorced from older urban areas and their propensity for social problems, pollution and ageing infrastructure. An unparalleled economic boom, resulting in high employment levels, rising living standards, mass consumption and car ownership promoted this urban expansion, together with an aggressive development industry and governments at different levels, which provided for public subsidies. What was at first a largely residential, sprawling and poorly serviced urban expansion has now become a fully-fledged deconcentration of almost all urban activities to the suburban fringe. In spite of the many economic, social and environmental problems associated with this type of development, it has remained by far the most popular form.

The Niagara Fruit Belt attracted the attention of preservationists, academics, politicians and various professionals very soon after the Second World War because its low-density suburban development was not only disorganised, discontinuous and poorly serviced but it was invading a finite area of unique agricultural land. Moreover, the small average size of fruit farms and the ability and willingness of farmers to sell all or part of their holdings as building lots set the stage for some of Canada's worse urban sprawl. The Ontario Government response was relatively swift but then became mired by inaction as the policy and land-use planning solutions were sought at the regional and local levels. It was to be 25 years later, after the worst of the suburban developments and sprawl had taken place, before there was a Niagara Regional Policy Plan in effect, and a further 20 years, to the present, before its restrictive measures regarding urban and rural development would be seriously put to the test.

This chapter will examine how such a unique agricultural resource, at times attracting considerable publicity and favourable responses from government regarding preservation, could be so constantly threatened and buried forever by urban development. In tracing the evolution of regional and local planning in Niagara over the last fifty years, it will be seen that the nominally very restrictive policies relating to where urban development can take place often mean very little in the face of a politically strong growth ethic, especially a Niagara-based one where a depressed economy has encouraged an 'any-growth-is-good-growth' philosophy. Moreover, a championing of local boosterism, and its local tax advantages, has come at the expense of acting regionally and diverting urban development away from the unique agricultural areas in favour of poorer areas in the southern part of the region. A booming agricultural and agritourism economy has done little so far to change political attitudes; and an Ontario Government which has been so busy devolving its planning responsibilities to the regional and local levels has provided little in the way of effective leadership, let alone binding legislation, on the issue of preserving the province's best agricultural lands.

The Early Role of the Provincial Government

The encroachment of urban development on unique agricultural lands in Niagara became a public issue and found its way into the media and the political realm in the early 1950s (Krueger, 1959). The previous, compact nature of urban areas, supported by the need to be near public transport services, was broken by the increasing ownership of the car and the ability to seek housing at a greater distance from jobs and services on cheaper rural land. In Niagara, it meant the explosive growth of housing, in a highly unplanned and sprawling fashion, in two critical Fruit Belt areas: to the east of Hamilton, in the former townships of Barton, Saltfleet and North Grimsby, and to the north and west of St. Catharines in the former townships of Grantham and Louth.[1]

Public pressure for the Ontario government to try to protect these lands for agriculture resulted in 1955 in an Ontario Department of Agriculture study of one particular municipality, the outcome of which was to assist both local and provincial governments in planning for this valuable resource (Irving, 1957). It was perhaps fortuitous that this study had, as a research assistant, someone who had been interested in Niagara since undergraduate days and who was to become an important advocate, lobbyist, and tireless researcher on behalf of agricultural land preservation in both Niagara and other areas of Canada: I refer here to the late Ralph Krueger, formerly Professor of Geography at the University of Waterloo Krueger went on to study the whole of the Niagara Fruit Belt as part of his doctoral studies at Indiana University, examining the physical factors necessary for fruit growing, the existing and future patterns of urbanization, and the basis for formulating a regional planning policy, including the redirection of urban development to lands less suitable for fruit growing (Krueger, 1959).

Continuing urban sprawl and little remedial action on the part of government attracted the attention of other professionals, including another Ontario government department, Treasury and Economics, concerned with the wider issue of the Niagara Escarpment, which was also under threat (Gertler, 1968).[2] Also, this same department employed another academic geographer to look at the social and economic impact of urbanization on farmers in the Niagara Fruit Belt (Reeds, 1969). Except for Metropolitan Toronto, there were no regional governments or planning authorities until the 1970s; and it was left to the Ontario Department of Treasury and Economics, in its 'Design for Development' program, to undertake regional planning across the province (Frankena and Scheffman, 1980). The program set up a number of regions, based on an amalgamation of existing counties, and set out on a large data collection exercise as a precursor for making major development plans for Ontario. Apart from various development concepts proposed for the Toronto-centred region, most of this top-down regional planning exercise barely rose above the data-collection phase, and gradually it was abandoned in favour of the bottom-up approach from local, and newly emerging, regional governments.

From the point of view of Niagara, there was the very real risk that its problem of the loss of a unique agricultural resource was being studied to death while a land-hungry form of urban development continued unabated. Yet another Ontario government department, Municipal Affairs, had joined in to help effect a solution. With its responsibility for local government administration, it set up a Commission to look at the reform of Niagara's municipalities; and one of the leading issues spurring on reform was poor land-use planning within the existing system (Niagara Region Local Government Review Commission, 1966).

During the 1950s, it had become clear that the existing structure of local government in Niagara was unable to handle the intricacies of extensive urban growth; and that urban sprawl and the needless loss of agricultural land resulted from poor or non-existent local planning (Gayler, 1979). Niagara, at this time, consisted of two counties (Lincoln and Welland) and 26 municipalities, varying in size from cities of 50,000 to large rural townships of a few thousand people and villages of a few hundred. The cities, including St. Catharines, Niagara Falls and Welland, were already fully developed within their boundaries, and the prospect of including further urban expansion rested on annexing parts, or the whole, of neighbouring municipalities. In the meantime, urban sprawl had leapfrogged city boundaries, and the rural townships, in particular, did not have the physical, financial and professional resources to cope with the problem. Indeed, it could be said that they exacerbated the problem. In order to improve their tax bases and be able to provide something in the way of urban services, these townships willingly went along with severing building lots and small subdivisions wherever any landowner requested it. All too often, severances of this nature were seen as a quick financial fix for the landowner, and rarely would township councils, which were more than likely made of the landowners' friends, stand in the way of these kinds of rights and freedoms regarding property usage.

The uncertainties and animosities caused by annexations resulted in St. Catharines, Niagara Falls and Welland, as well as some smaller municipalities, approving some form of regional government for Niagara and petitioning the Ontario government to bring about change (Niagara Region Local Government Review Commission, 1966). The Commission Report recommended a two-tier system of Niagara Regional government and 12 local municipalities, focusing on the 10 largest cities and towns and two predominantly rural townships, and effectively removing the competition between land-deficit towns and cities and neighbouring land-surplus townships. The problems of land-use planning that had long been identified were to be solved in the context of local and regional planning under the authority of the 1946 Ontario Planning Act, rather than any top-down decision-making from the Ontario government. The earlier local enthusiasm for regional government did not last long,[3] but the Ontario government overrode local opposition and endorsed virtually all the recommendations of the Commission (Government of Ontario, 1969; Krushelnicki, 1994). The Regional Municipality of

Niagara and its 12 local municipalities came into being on January 1, 1970, and among its first actions was the legal directive to establish a Regional Plan by December 1973.

The Failure of Regional Planning, 1970-1981

The course of action to do something about the Niagara Fruit Belt's land-use problems, as opposed to studying and talking about them endlessly, as had been the case for the previous 20 years, became a controversy of major proportions between Regional and local governments on the one hand and a public-interest group, the Preservation of Agricultural Lands Society (PALS), and at times the Ontario government on the other. It was to lead to a 11-year test of wills, an Ontario government throwing up its hands in favour of its quasi-judicial appeals court on planning matters, the Ontario Municipal Board (OMB), leading to one of the longest and most bitter OMB hearings in memory to bring closure to the first Regional Plan (Gayler, 1979; 1982a; 1982b; Jackson, 1982; Krueger, 1982).

Clearly, talk in the past was cheap; but when it came to deciding action in terms of an approved Official Plan, especially the designation of urban-area boundaries on a map, then the prospect (or otherwise) of realising millions of dollars one day in selling developable land became a critical matter. It soon became clear that there would be no agreement as to the amount of land designated for urban purposes, nor its location. Also, rural areas in Niagara would come under the same planning rules as urban areas. No longer would it be open season in the countryside for any type of urban development. The Ontario government had meanwhile restricted the worst of unserviced subdivision development; now it was the turn of Regional government to plan what could go where. In the Niagara case, for example, attempts to restrict development on the unique agricultural lands in the Fruit Belt were hardly new; but they met with opposition from landowners and developers when the reality of lost millions of dollars and restricted future choices began to sink in.

Both the urban and rural land-use policies were not ready by the December 1973 deadline. Many of Niagara's towns and cities already had approved or proposed local plans, and Regional attempts to make amendments ran into opposition whenever potential financial losses would result. In the end, the Region largely endorsed the wishes of the towns and cities. However, in doing this it took no account of the actual need for future urban land (Gayler, 1982a). The local municipalities' earlier proposals were largely a wish list, and when aggregated by the Region became a positive embarrassment. The proposed urban areas forwarded to the Ontario government in October 1974 could accommodate a population of 640,000; but the population projected for 1991 was only 510,000. The former figure was probably low because it took no account of raising densities or development that had already been approved outside the urban-area boundaries. Meanwhile, the population projection continued to fall as new census and assessment data became available; by 1977, a figure of 415,000 was proposed for Niagara in 1996. As it happens, this estimate was not too far wrong 20 years later.

This wish list made a mockery of any Regional Plan. First, the prospects of urban development and financial gains were unlikely to be realised by large numbers of landowners, at least in the short and medium term. However, no individual landowner wanted to give up on the chance of development, which placed uncertainties on farm operations and increased speculation. Secondly, too much designated urban land ran the risk of over-extended servicing and unnecessary public expenditures. Thirdly, the Region's proposed commitment to preserve unique agricultural land in the Niagara Fruit Belt and redirect development to poorer lands south of the Niagara Escarpment somehow had a very hollow ring to it.

The period from 1974 to 1981 was fraught with regional and local intransigence, provincial government intervention, and the Regional Plan's urban and rural land-use policies in effect being decided by the OMB. Rather than allow professional planners to designate urban areas in accordance with need and stated policy goals, each local municipality clamoured for as much land as possible. This overprovision of urban land and the threat in some municipalities to unique agricultural land eventually resulted in public pressure for the Ontario government to intervene. Fortunately, it became an issue in the 1975 provincial election, a time when Ontario's Progressive Conservative government was already pressured on the issue of regional government in general and faced the prospect of losing power for the first time in over 30 years. The Ontario Ministry of Housing took this opportunity to reject the proposed urban areas for Niagara and sent the matter back for revision. But the Regional and local politicians and their landowning and developer friends were not to be dissuaded: the first revision only removed approximately 500 acres from the 26,000 acres proposed for development, and the Ministry of Housing responded by arbitrarily removing another 3,000 acres in the critical Fruit Belt area. The latter decision could be regarded as 'smoke and mirrors' since some of the areas removed were already built upon.

The difficulties that such a cat-and-mouse game made for a Progressive Conservative government and its likely supporters at the local level resulted in the matter being eventually handed to the OMB and thus taken out of the political arena. Between October 1978 and November 1980, there were two hearings, one a year in length, involving eight of the twelve local municipalities and all five municipalities in the Fruit Belt. The hearings can be described as a David and Goliath affair, with all Regional and local governments, together with landowners and developers and taxpayer-supported professional and legal talent, defending the Regional Policy Plan, while the opposition was a poorly-financed, public-interest group (PALS), dependent on the charity of professionals, including Ralph Krueger and other academics, sponsored walkathons and bake-sales and a helping hand from Ontario's Legal Aid fund. The lands in dispute included not just those within the proposed urban-area boundaries, but landowners and developers just outside the boundaries were ever hopeful that their lands might be included in an urban area as the result of an appeal. The land being questioned was literally attacked or defended lot by lot, in a process that may have been fair to the individual landowner but was certainly designed to wear down a group such as PALS,[4] as well as to be found wanting in terms of overall Regional policy. The Ontario government played virtually no role during the hearing process; although it formulated provincial guidelines for the protection of different quality agricultural lands (Government of Ontario, 1978), it was left to PALS to defend these.

The OMB, in its decision in 1981, paid tribute to the work of PALS in being the sole defender of the wider public interest in preserving unique agricultural resources in Niagara (Jackson, 1982). However, the OMB recognised that although one third of the land proposed for development in 1974 had by now been removed, there were still excessive amounts of land available. These surplus lands in the Niagara Fruit Belt were defended by the OMB on the basis of allowing local municipalities sufficient time to adjust to these urban-area boundaries being considered permanent, as well as future urban development being redirected on to poorer lands.[5] Likewise, it gave farmers greater security in future planning, and was seen to be fair to all local municipalities; although factors such as location and accessibility could not help but favour some municipalities over others in terms of where future development would take place.

Outside the urban-area boundaries the OMB formulated rural land-use policies based on the quality of land base with the good tender-fruit and good grape land areas being restricted to agriculturally-related developments. These planning policies in rural areas have been viewed by most local residents as a new departure and Draconian; but while they certainly have restricted extensive, unplanned urban sprawl, there have always been opportunities for urban development in the countryside, i.e. for a slower demise of an agricultural resource by attrition.

First, it has never been clearly argued what is meant by agriculturally-related; and to add to the uncertainty, a number of urban uses have been allowed in the Fruit Belt because they are deemed to be compatible with agriculture, including churches and their car-parks, cemeteries, social clubs and sports' facilities. Again, it is not clear what is meant by compatible, especially when the urban use results in the land being lost to agriculture forever.

Secondly, Regional planners were not sure how many severances for building lots had been granted before 1981 and had not yet been developed. And to add to this uncertainty is the potential for urban housing arising from the continual splitting or amalgamation of farm holdings as well as infilling.

Thirdly, Niagara has never come to terms with the loophole that is created by granting retirement severances. This grants a farmer who has been in full-time farming for 20 years the right to sever a building lot on the farm property on retirement. But there is no requirement for the farmer to actually build a house rather than sell the lot. Furthermore, it has to be recognised that one day the farmer or his/her heirs will sell the house anyway, and at that time it could easily become yet another piece of ex-urban real estate. The OMB hearing officers excluded retirement severances from the Regional Policy Plan, but the Region of Niagara appealed this decision to the Ontario Cabinet, who allowed it providing there had been full-time farming for 20 years prior to December 1973, the date by which the Policy Plan was to have been completed.

Fourthly, there is one threat to this unique agricultural resource that no amount of planning or government legislation can prevent, simply because it would be unconstitutional: there is nothing to stop an urbanite intent on living in the countryside (and yet stopped from subdividing land) from buying up the whole farm and either renting some of the land to a bone-fide farmer or, worse, converting all the land to non-agricultural uses such as lawns, rough pasture, meadow, horse paddocks and wood-lots. The Niagara Fruit Belt's propensity for small farms, and a past record of economic difficulties for many of these farmers, only heightens the potential for these rural-estate and hobby-farm conversions.

The Regional Policy Plan in Operation, 1981-1995

The Regional Policy Plan could be viewed as a document that offered, for the first time, very strict limits in what unique agricultural lands in the Niagara Fruit Belt could be used for (Regional Municipality of Niagara, 1988). In practice, however, these restrictions are quite illusory. Regional and local politicians have never been convinced by the idea of the permanency of urban-area boundaries; the redirection of urban development south of the Niagara Escarpment has largely remained a planning concept rather than a serious plan of action; any exception to these restrictive rules can be argued on the basis of an Official Plan amendment, which can be easy to achieve where there is a sympathetic political audience listening to plausible arguments, on behalf of a client, from lawyers and planning consultants; moreover, if a land-use change can be viewed as helping a farmer in economic difficulty, then the likelihood of success is that much greater.

The excessive amounts of unique agricultural land were designated by the OMB in 1981 have been borne out by the fact that for the next 15 years there were no major changes to the urban-area boundaries. Apart from Grimsby, and the pressures of suburbanisation eastwards from Hamilton, there have been no serious proposals; and the minor changes often reflected changing land classifications and clearing up a variety of anomalies. In so many respects, unique agricultural land has been protected, not by a Regional Policy Plan and sympathetic landowners and politicians but by a Regional economy that was both faltering and making fewer demands on land conversion.

Since the 1960s, population growth in the Niagara Region has consistently trailed that of Ontario and Canada as a whole, and especially much of the rest of the Golden Horseshoe around Toronto (Gayler, 1994). Between 1976 and 1986, in fact, Niagara grew by a derisory 1.3 per cent. This status quo position reflects a slowing of population increase in the so-called echo generation, compounded by high losses of manufacturing employment in both the traditional heavy industries and the newer ones such as transportation, and Niagara's failure to attract so many of the newer industries and services. Meanwhile it continues to have a net loss of its young people and has become more attractive as a retirement area. In spite of favourable house prices, attempts to promote Niagara as an outer suburbia to Metropolitan Toronto wither after one considers the cost and social disruptions of an approximate four-hour daily commute.

While there has been no large-scale clamouring for more urban-designated land, there has been sufficient activity taking place over this period, that regional and local politicians readily supported, to indicate that had there been any major proposal to extend urban-area boundaries, it would likely have been welcomed. It is clear that lip service is paid to the preservation of valuable agricultural land for when a developer or landowner comes forward with a non-agriculturally-related proposal, it is so often supported because the merits of the case outweigh the agricultural considerations. The evidence for this comes first from the various schemes for wholesale land conversions during the early 1990s and second the steady attrition of rural land for urban-related purposes.

The schemes for wholesale land conversions were an outgrowth of the difficulties being faced by tender-fruit farmers in Niagara and an attempt to use the planning process to solve their economic woes. In the early 1990s, the effects of the new NAFTA regulations opened up the Canadian tender-fruit market to unwelcome American competition; the economy meanwhile was in the throws of a recession; and there was a strong likelihood that existing government financial incentives to farmers might be scaled back. These economic difficulties happen to coincide with the ten-year review of the Regional Policy Plan, and many politicians and farmers saw a chance to amend planning policies in order to help those in difficulty (Gayler, 1996).

For the most part, the review looked like being a painless exercise, designed to take care of where urban development would take place once existing developable lands were exhausted. However, the public meetings were hijacked by an angry minority of the 900 or so tender-fruit farmers, pleading for a relaxation of the strict policies relating to the severing of lots for non-agricultural purposes. Regional politicians succumbed to the pressure and approved Amendment 60 to the Policy Plan, no doubt against the wishes of the professional planner, giving the right to financially hard-pressed, full-time, tender-fruit farmers to sever one-acre lots from their farms; the exact number of severances depended on the size of the farm, and could be to a maximum of seven (Regional Municipality of Niagara, 1991). Official Plan amendments have to be approved by the Ontario government; and this blatant invitation to promote urban sprawl by allowing hundreds of building lots throughout the Niagara Fruit Belt, as well as use the planning process to further the financial interests of a particular group, met a road-block with both the Ontario government and its civil service. Also, only the government had the power to send the matter to the OMB, and in spite of local calls to do so, sanity prevailed and Amendment 60 did not become subject to an appeal.

The desperation of some local tender-fruit farmers was such that the pressure on Regional and local politicians was kept up that most of, and a new avenue to attract the Ontario government's attention was sought. It was not long in coming. The new NDP government in Ontario, elected in 1990, was keen to bring about reform to both the planning process and the underlying philosophy behind planning, and appointed a Royal Commission under the Chair of John Sewell, a former Mayor of Toronto, to undertake a study into the deficiencies of the existing Planning Act. Part of that study was a series of public hearings in various parts of the province; and needless to say when the Commission came to Niagara, irate and frustrated tender-fruit farmers dominated the meetings, pleading with the commissioners to relax the restrictions relating to urban uses on unique agricultural land. At the time, the firm yet compassionate reply was given that there could be no guarantees that planning could be used to help a particular group in need. Left unsaid was the obvious retort that what was being proposed as a means of helping the farmer was in essence Amendment 60 and the return to urban sprawl. The Commission's Report, and the subsequent Bill 163, did quite the opposite: more efficient urban development, using less land and encouraging a better use of infrastructure and public service facilities, was clearly stated, and any urban expansion would have to be justified in terms of the quality of agricultural land (Commission on Planning and Development Reform in Ontario, 1993). Furthermore, local plans would have to be 'consistent with' these policy statements, rather than the former, and looser, 'have regard to' such policies.

While the tender-fruit farmers were thwarted again, their key argument that saving agricultural land was destined to fail if one was unable to save the farmer first, was not lost on the Ontario government, and various options were explored by the two levels of government during 1994-1995 to help farmers in economic difficulty to stay on the land. The option eventually chosen was agricultural easements, where a farmer would be given a one-time subsidy, based on the number of acres that were eligible, in return for amending the title to the land to exclude any urban development in perpetuity (Daniels, 1991; Regional Municipality of Niagara, 1995a). However, such payments were in no way tied to agricultural improvements or change; and the question was asked, what would happen when the money ran out and the farmer was still in economic difficulty. Would there be renewed pressure for some urban change? While everyone gave the nod to spend taxpayers' money in this way, the magnitude of the problem in Niagara, as well as a recession and mounting public debt, threatened to scuttle the program. The prospect of 10,000 acres of tender-fruit land that could be included, and a subsidy of approximately $10,000 per acre, meant a government pay-out of about $100 million. In the end, this program was cut back to $20 million, part of which would be locally financed, and only 2,000 acres would be eligible; the subsidy would be spread over five years, and the farmers chosen would be dependent on the agricultural value of their land and the potential threat of conversion to urban purposes.

There now remained the potential for conflict within the agricultural community as to who received a subsidy and who did not. However, the program brought a measure of peace to the Niagara Fruit Belt, and diametrically-opposed forces such as PALS and the agricultural community agreed to work together on the matter. The program also sent a clear message that no longer were farmers having to bear the brunt of preserving unique agricultural land; instead, what had been regarded as a public demand for such preservation was now being translated into a public responsibility with a widely shared financial commitment. Together with a much more restrictive Planning Act, it seemed as if the defence of Niagara's unique agricultural lands was more assured than it had ever been. Furthermore, the major part of the ten-year review of the Regional Policy Plan had been the establishment in practice of the earlier policy encouraging the redirection of future urban development on poorer lands to the south of the Niagara Escarpment (Gayler, 1991). Although no actual location was chosen, it was agreed by Regional Niagara Council that the area between Niagara Falls, Welland and Thorold, close to Highway 406, would contain future urban development, with highway commercial development focusing on the Queen Elizabeth Highway south of Niagara Falls.

The surprise election of a Progressive Conservative government in Ontario in June 1995 resulted in a marked change in the way that agricultural, urban development and planning issues were dealt with in the province. The new government's Common Sense Revolution was a commitment to slash public expenditures in order to reduce public debt, create a better business climate and reduce personal income taxes; and one of its first actions in July was to abolish hundreds of programs, including the agricultural easement program. The uncertainties created in the agricultural community were accentuated by reversals to the previous government's Planning Act. The new Bill 20 allowed planning decisions by local and regional municipalities to once again 'have regard to' rather than 'be consistent with' provincial policies (Government of Ontario, 1996). Also, these policy statements were revised in order to remove some of the previous restrictions on urban development. While it was still considered preferable to have urban expansion on lower quality land, there was no longer any restriction on expanding on high quality land or specialty crop lands if alternatives were not available.

Niagara's politicians and farmers were not slow in coming up with an alternative to agricultural easements. Eyeing a sympathetic government caught up in a cash-flow problem, they set out to present proposals that would generate money for farmers at no cost to government. The Region formed an Agricultural Land Use Subcommittee, consisting of major stakeholders in the agricultural community, Regional councillors and members of the public, with terms of reference to set a course of action for the next five years and to present its findings to the Region's Planning Services Committee in less than six weeks. The resulting document was based on four meetings, no public hearings and virtually no input from professional staff (Regional Municipality of Niagara, 1995b). As one Regional planner (who must remain anonymous) said, 'it (the document) was a political report not a planning one.' Indeed, the usual array of signatures from junior planner to CAO were conspicuously absent from the document when it was presented to the Planning Services Committee.

The Subcommittee's work had a very familiar ring to it, and certainly needed little time in preparation. In part, it was Amendment 60 revisited, but expanded to include all farmers in the Region whether in economic difficulty or not. Depending on the size of the farm, up to three severances could be granted.

Eligibility for a retirement lot would be changed from farming continuously since before 1973 to only the last ten years. Downsizing farms into two or more parcels would be changed from a dependence on economic viability to simply minimum acreages. The restrictions relating to infill lots were relaxed to permit all farmers an opportunity to comply. Severing surplus farm dwellings was originally allowed only with the amalgamation of adjacent farms; now it was to be extended for whatever the reason for initial construction. Finally, the uses permitted in agricultural areas were to be relaxed in favour of a variety of non-agriculturally related activities that could use existing farm buildings and assist farm incomes. Again, the Official Plan amendment needed provincial government approval, and while these local initiatives would likely be more favoured by a Progressive Conservative than an NDP government, they were perhaps too farcical for any responsible provincial government to entertain. The amendment was denied.

Post 1995: Regional Planning in a Period of Economic Upswing

After a period of relative calm, the last half of the 1990s has seen a number of changes in both the Niagara economy and attitudes towards urban and rural planning which again have heightened concerns for containing urban land uses and preserving unique agricultural land. In spite of a Regional Policy Plan having very restrictive statements about what types of urban development should take place where, the Policy Plan is in no way cast in stone; and one is left to wonder, therefore, how far we have come in 50 years with respect to preventing the constant attrition of our best agricultural lands. While many can appreciate the problem, and know what needs to be done (and have known that for a long time), there is often little political will at the regional and local level to do anything about it, resulting in conflict and uncertainty.

Niagara's Amendment 60 (Revised Agricultural Policies) was eventually watered down into a very tame document, and the prospect of urban sprawl on a grand scale with thousands of building plots spread through the Niagara countryside disappeared altogether (Regional Municipality of Niagara, 1997). At least on this aspect of development we have learned from past excesses, but on most of the other aspects, considered below, we certainly have not.

Urban Land Uses in the Countryside

It continues to be open-season in the countryside for a large number of urban land uses, serving a predominantly urban population, that either cannot be contained in the city, or for reasons of comparative land costs are preferred in the countryside. In most instances, it is the owner of the rural land who is behind the proposal for change, and often it may make better economic sense than farming the land. The quality of the agricultural land tends not to matter; and except for the City of St. Catharines, most local councils, and also Regional council, will give permission for the land-use change. St. Catharines has always had a stricter policy regarding the types of change that are entertained in its rural areas, and has often been accused of being an urban-oriented council that does not understand agricultural needs.

There are any number of convincing reasons why the quality of agricultural land does not matter and permission should be given for a zoning or Official Plan amendment change. Politicians are still won over by the landowner who is 1) not able to make a success of farming even though neighbouring farmers on the same quality of land are successful, or 2) can argue that the quality of the land is not as good as the Policy Plan suggests. In many cases the land has been held speculatively and allowed to deteriorate in order to bolster the argument that it is not good quality land. It comes back to the age-old argument, why save agricultural land when you cannot save the farmer; however, perhaps we should be insisting upon, in first instance, a change of ownership, rather than land use, to see if someone else can do a better job of farming the land. Invariably the plight of the individual landowner carries more clout than the wider public good of saving unique agricultural land for future generations. The former is known to regional and local politicians, or, if not, a good planning consultant and or lawyer can make politicians aware; on the other hand, the latter is a more nebulous concept, clearly blocking change and by some people's definition – progress, offering little in the way of financial gain today, and damned as the sort of argument that urban intellectuals, and especially members of PALS who do understand the agricultural community, are likely to come up with.

A second reason for land use change is the conflict with neighbouring uses, usually urban-related, which, it can be argued, make for increasing difficulties in running an agricultural operation. This rural-urban interface argument favouring an urban-related change means an ever-increasing urban sprawl, since development has long been more closely tied to highway locations, accessibility, and individual decision-making, than any notions of contiguity with existing built-up areas. Conflicts could be reduced in the first instance by effective buffers between urban and rural uses. However, one suspects this does not happen because the conflict is being used to promote the change of use in the first place. Even where the farmer wants to continue in operation, and buffers are tried, adjacent urban land uses can affect what the farmer would like to do, hastening perhaps further urban change.

In spite of the resolve of most farmers to stay on the land, and of politicians for wanting that to continue, urban changes in the countryside are viewed as part of rural development and afford greater economic and social opportunities for the local population. These include the increase in the rural tax base, attracting urban tourists to spend money there, diversifying employment opportunities, and increasing rural incomes. However, the type of development that can come, and its form, can vary widely in terms of acceptability. Perhaps the most acceptable is where the non-agricultural activity can be incorporated on the farm, increasing and stabilising farm income and providing employment for a family member without having to leave home; this is of great benefit where that person also helps with the farm operation. Most rural plans allow for this multi-use function so long as the various uses are compatible.

Most urban uses, however, are divorced from agriculture and come with a retinue of problems. Almost all uses take agricultural land out of production for all time. Some, particularly golf courses, driving ranges and sports' fields, are said to offer the opportunity of a return to agriculture; however, there are no examples of this happening, and probably the level of capitalisation and physical change, e.g. the use of pesticides and fertilisers on golf courses, would inhibit any return. Many land-use changes are argued on the basis that they are agriculturally-related, and therefore allowed for in the Regional Policy Plan. Even these changes are not without their problems and will be viewed under agritourism later.

The next category of land use is often argued before Regional and local councils on the basis of it being compatible with agriculture, in the sense that no buffers are needed between rural and urban uses and the operation of the one is not likely to harm the other. Many of these urban uses are low intensity, and not large generators of traffic on a regular basis. Included here are churches, cemeteries, schools, and social clubs with adjacent sports' facilities. While the compatibility rider may apply, many of these uses take up a large amount of agricultural land, especially for car parking. The growth of new churches, or the splintering of old ones, has been of particular concern in the Niagara Fruit Belt. The siting of these churches is questioned when most of the congregation is likely driving out from nearby urban areas; but the lower land costs in rural areas and the even better prospect of the land being already owned by or donated to the church help to sway the politicians.

The compatibility issue is frequently a ploy to affect a zoning or Official Plan amendment change; but in so many instances, urban and rural land uses do unduly influence one another and demonstrate that there are two very different and conflicting cultures at work. To introduce more urban uses is to cast the shadow effect more widely and inhibit agriculture further. Moreover, in the Niagara Fruit Belt, the extent of urban development beyond the urban-area boundaries is such that we are seeing agricultural changes being opposed by a rural, non-farm and perhaps ex-urban population. Niagara's rural residential population finds country living compatible so long as the reality of adjacent agricultural land use is the quiet, romantic and bucolic view of orchard, vineyard and field crops. However, the introduction of noise, e.g. bird-bangers and agricultural machinery, smells, e.g. fertilisers and intensive animal operations, lights, e.g. from greenhouses at night, inconveniences, e.g. mud left on the road from a tractor, and obstruction of the view, all of which may be very necessary in the efficient running of the farm, changes that reality and intensifies conflict between the two groups. The recent rapid growth of Niagara's greenhouse industry is an example of this conflict (Gayler, 2001a). Rural residents, who had previously bought into the views of open countryside, found an expanding greenhouse operation not many yards from their front windows. Since greenhouse operations are zoned agricultural, even though the land underneath may not be used, permission has to be granted for these developments even on the most valuable of Niagara's agricultural lands.

The final group of urban uses are those that either consume large amounts of rural land or have serious implications in terms of the escalation of development in the future. Two examples in the Niagara Fruit Belt, the Seventh Street Autoplex/Farmer's Market in rural west St. Catharines and a Culinary Centre in the Town of Lincoln, both of which failed, at least in their original form, clearly attempted to take advantage of site characteristics respectively, freeway intersection, and Niagara Escarpment scenery and proximity to the owner's winery; but they were not categorised as agriculturally-related uses and could just as well have been contained in an urban area. The 35-acre autoplex was a blatant attempt to reorganise a number of St. Catharines car dealerships in a more convenient location with room for recreational and eating facilities (Regional Municipality of Niagara, 1992). When that failed to win approval, a more compatible 100,000 sq. ft. farmer's market was tried (Miller O'Dell Planning Associates, 1997). But this year-round operation was an, albeit tasteful, urban superstore in the making, and there were concerns about servicing and the expansion of urban-area boundaries; eventually, it was pared down and approved as a 25,000 sq. ft. seasonal farmer's market and winery. The Culinary Centre was a bold move and associated with a nearby winery and an expansion in agrotourism (Regional Municipality of Niagara, 1998). The proposals ran into opposition from public interest groups and professional planners because of the 55 chalets for accommodation. This is something that is contrary to rural policies, where only small bed and breakfast operations are permitted, has aesthetic implications being on the bench of the Niagara Escarpment, creates servicing problems and could all too easily be converted to rural residences. However, the tourism and other economic advantages swayed not only the Regional and local Councils to approve the Centre, against staff advice, but also the Niagara Escarpment Commission. The Centre was in the Commission's planning area, and while this development was also contrary to its own plan, an increasingly pro-development Commission allowed it. Commission decisions, unlike those of the OMB, can still be appealed to the Ontario Cabinet, and in a surprise move by a pro-development government, the Cabinet rejected the plan.

The few, large, bold moves in a sensitive agricultural area certainly attract media attention and grandstanding by the parties concerned, both for and against, but in many ways it is the constant attrition of small parcels of land for whatever urban purpose that can go almost unnoticed and yet have just as damaging long-term consequences on agricultural operations. Moreover, many of them do not even contravene rural policies. For example, as mentioned earlier, while rural estate development may not be allowed, the problem can easily be solved by buying up one's own farm. Secondly, Niagara has been allowed by the Ontario government to change the rules on retirement severances. Originally, one had to have farmed continuously for 20 years prior to 1973, thus ensuring that the problem of retirement lots becoming building lots for ex-urbanites would eventually die out; but in an attempt to be fair to younger generations of farmers, by having the rule changed to any twenty-year period of full-time farming, the Ontario government has ensured that one aspect of urban sprawl will persist.[6]

The Expansion of the Urban Area Boundaries

The second, and far less invidious method of urban growth is the contiguous expansion of urban areas. The principal concern perhaps is where this will take place, especially since permanent urban-area boundaries for Niagara Fruit Belt municipalities, which were suggested in the 1981 Regional Policy Plan, were never put to the test and were eventually abolished. The OMB decision of 1981 allowed for generous boundaries in all municipalities, but at some time the status quo position would be challenged as developable land ran out. Redirection of urban development was in the original Policy Plan; but it was not until the ten-year review in the early 1990s that a general area was chosen, and not until the late 1990s that the area of Port Robinson West, just to the north of Welland and adjacent to Highway 406, was chosen as the site for major urban expansion in Niagara (Regional Municipality of Niagara, 1999a).

In a Niagara that acts more like 12 individual municipalities than an integrated region, the question arises how other municipalities will react to the loss of the status quo and the chance of major expansion of their urban area. There have already been minor challenges to the urban-area boundaries in Grimsby, Lincoln and Niagara-on-the-Lake, and no doubt the heat will be raised as developable land runs out in the next 10 years. But the success of a planned 550-acre expansion to the urban-area of Fonthill, in the Town of Pelham, in 2000, has effectively shot down the policy of redirection of urban development on to poorer quality lands (Ontario Municipal Board 2000; Regional Municipality of Niagara, 1999b).

Fonthill, which has grown from an agricultural village to a bland, could-be-anywhere suburb over the last 30 years, has become a highly desirable middle-class residential location. Unfortunately, it is situated in the centre of some very sensitive natural areas and unique agricultural lands, and the only area south of the Niagara Escarpment that is akin to the Niagara Fruit Belt to the north in terms of soil, microclimatic conditions and land use. The region council chose to back the municipal boosterism of the Pelham council, in spite of all the doubts of professional staff who could recognise the predominant land-use classification of good tender-fruit lands, the restrictive policies of the Regional Policy Plan, and the in-your-face evidence from the main highway of existing land-use as a successful nursery operation. PALS appealed to the OMB to reverse the Regional and local decisions; but the sheer weight of professional advice and evidence bought by government and developers, compared to that which could be afforded by PALS, and the conflicting nature of some of that evidence, made for another David and Goliath situation.

It was successfully argued by professionals representing the pro-development forces that in spite of the evidence of previous or present use as tender-fruit and nursery lands, the agricultural capability of these lands was not as valuable as previously thought or as designated in the Regional Policy Plan. Besides, it was argued that better lands existed elsewhere in the Niagara Fruit Belt and these lands adjacent to Fonthill were already subject to various rural-urban conflicts that made agricultural production difficult. Although not introduced as evidence, this statement could easily apply almost anywhere in the Niagara Fruit Belt! Moreover, provincial policy statements point to communities having a 20 year supply of residential land, which the proposal would allow for, as well as permit urban areas to expand on to valuable agricultural lands if the need can be demonstrated. The OMB, in dismissing the appeal, argued, *inter alia*, that communities such as Fonthill are mandated to provide for future growth under the Ontario Planning Act and policy statements at all three levels of government.

A regional policy of redirection, whereby some communities would not expand, at least spatially, in favour of other communities that would, can therefore be seen to be of dubious value unless there is a fundamental policy change at the provincial level. It will be seen later than even in St. Catharines, where the city council is on record favouring permanent urban-area boundaries and redirection of development to other communities, these issues remain active, especially among landowners and developers, and troublesome for the future of a unique agricultural resource.

Agricultural Change and Land Preservation

During the 1990s, there have been some attitudinal shifts with respect to urban containment and the preservation of agricultural land that could indicate that the Niagara Fruit Belt will be better protected in the near future. The long-held views that farmers in economic need should be helped by land conversion for urban purposes, and there is little value in saving the land if one cannot save the farmer, are being challenged in new and essentially more powerful ways; also, these ways relate to bringing economic diversity and greater financial returns to the agricultural community, especially in the Niagara Fruit Belt. No longer is it simply preserving agricultural land at the farmers' expense for the enjoyment of urbanites on a Sunday afternoon drive or ex-urbanites who live there.

The key to the change has been important changes in the agricultural industry itself, and specifically related to the grape and wine, horticultural and tender-fruit industries. Increasingly, since these industries will only prosper if the land base is protected, then it is in their interests to join forces with groups such as PALS and other public and professional groups to fight for the protection of unique agricultural land for all time. Furthermore, it has long been recognised that it is only in the Niagara Fruit Belt and one or two other small areas in Canada that this protection is relevant. Although in the interests of expanding the land base for these crops, and reducing intra-industry competition for land, research is being undertaken at this time to test the capabilities of other areas.

The principal change stems from developments in Canada's grape and wine industry and the spin-off effects in agrotourism (Gayler, 2001b). After almost a century of languishing with an inferior and constantly panned product and temperance-inspired behaviour, the grape and wine industry began a major shake-out in the 1970s. French hybrid and vinifera grapes were introduced and quickly took over in wine-making from the local Lambrusca varieties; and the notions of a wine culture, a quality product with its climate-based good and bad years and the artistry of the winemaker took over from the factory-based, Fordist, industrial production. The industry saw the closure, take-over or rationalisation of the older national firms and the establishment, so far, of over 40 small estate wineries in the Niagara Fruit Belt. It has been a hard thirty years as the Canadian industry has had to fight a series of 'demons'. These include foreign competition especially that resulting from NAFTA and the easier access of U.S. producers;[7] the hindrance of economies of scale because of the size of the industry and thus the higher production costs than elsewhere; unsympathetic federal and provincial governments, which allow, for example, lax rules on the importation of wine from beyond North America, and for foreign wines to be blended with Canadian ones and still be marketed as Canadian wine; a Canadian society more closely attuned to beer and spirits that has to be educated about not just a new product but almost a new life-style; Canadian restauranteurs who are still loathe to have Niagara wine products on their menus; and a Canadian public, unique among wine-producing countries, who still consumes more imported wine, reflecting not just the small Canadian production but the greater snob appeal of the European product.

However, growth and innovation in Niagara have recently been related to the promotion of a superior product, recognising a single grape variety, its quality in a given year, and grown perhaps in a particular Niagara vineyard. The industry's VQA (Vintner's Quality Control) designation of its superior wines and the government's endorsement of this process have helped to promote the product on the international scene. From once being universally panned, various Niagara wineries have been awarded top prizes in international, blind-taste competitions. The crowning achievement perhaps for Niagara producers, with important implications for the future, was the recent permission to market their ice-wines in France.

Of equal importance locally has been the further promotion of the grape and wine industry through agrotourism developments. Estate wineries are approved on the basis of manufacturing plant, warehouses and distribution facilities that are situated amid a certain acreage of vineyards and thus fairly dispersed locations across the Niagara Fruit Belt; but all of them have been allowed various tourist facilities, including winery tours, wine tasting, retail sales and gift shops, restaurants, banquet and conference facilities, bed and breakfast accommodations, and an ever-expanding and imaginative, year-round, events calendar where an occasion (e.g. Shakespeare in the Vineyard, Jazz, art and antique car exhibits, ethnic foods, wine-related occupational training, and recreation) is tied to the less than subtle promotion of that winery's product. Niagara Falls already attracts ten million visitors per year, and Niagara-on-the-Lake about three million, so a ready market is there to be tapped. However, agrotourism and the wine experience are being promoted by the private and public sectors as a single-purpose, up-market and sophisticated, tourist experience among the urban upper-middle class in southern Ontario and western New York.

An important aspect to this industry is the marketing of its aesthetically-pleasing rural locations, dubbed by many, Napa Valley North; and whether from the point of view of wine-making or agrotourism, there is a strong entrepreneurial interest in the responsible use of the agricultural land resource and the protection of those lands from urban encroachment. In ten years, we have come from an agricultural industry that was indifferent or hostile to protecting that resource at the expense of a farmer's right to a fair income, to one where winery owners are at the forefront of lobbying all levels of government, as well as other members of the agricultural community, to elicit better controls over land conversion. If their industry is to exist, let alone expand, their land base must be protected; and since most urban land uses can afford to pay more for the land than the grape and wine industry, those controls must be quite rigid, and moreover, enforced.

The tender-fruit industry has likewise gone through a renaissance in the 1990s. From an industry hit hard by Free Trade and a recession, and readily promoting urban sprawl in order to get a quick economic fix, there has recently been a resurgence of demand for fruit products, as well as a projected need for more tender-fruit land to meet growing demand (Ontario Tender Fruit Producers' Marketing Board, 1999), and no large-scale pandering anymore for severing thousands of building lots.[8] The tender-fruit and grape and wine industries overlap such that agrotourism interests of the one also serve the other. Farmers' markets and farm-gate sales, which similarly are expanding, are found along the same main roads in the Niagara Fruit Belt as the famed Wine Route that connects all of Niagara's wineries.

The third part to this successful trinity is horticulture and the greenhouse industry. In the last ten years the area under glass has more than doubled and it remains by far Niagara's most valuable agricultural sector with a robust future (Gayler, 2001a). With one notable exception, it is an industry that has shied away from on-site retail sales and attracting tourists; but it mirrors the other two industries in that it is tied very largely to the Niagara Fruit Belt area, suffers more than the others from rural-urban conflicts, and is just as keen to protect the agricultural land base. Unfortunately, the greenhouse industry is engendering a good deal of uneasiness among its tender-fruit and grape farming neighbours; and it is often argued, although not altogether successfully, that since greenhouses rarely use the soil on which they are located, they could just as well be redirected to poorer soils south of the Niagara Escarpment.

The success of these three agricultural industries is perhaps the best solution possible for the preservation of Niagara's unique agricultural lands, although that success may result in there being not enough land available, at the right price and in the right configuration, in the Niagara Fruit Belt. Also, the continuing growth of agrotourism unfortunately brings about its own pressures. Developments at wineries, including not only the various industrial and associated buildings but also the various tourist facilities, car parking and landscaping, take away from the adjacent vineyards. And the legacy of urban sprawl is such that wineries, however well designed they may be in their pseudo-French chateau or neo-Georgian colonial look, often conflict with the views and privacy considerations of rural, non-farm neighbours. Also, more tourists mean more traffic, leading to necessary road changes for safety and capacity reasons; and more traffic results in spin-off activities that tourists might need e.g. a garage, ice-cream store, antique barn, recreational activities, and fast-food outlets, most of which will take land out of agricultural production. The irony is that agricultural land is being lost whether the rural economy booms or not, and begs the question, can much of this agrotourism and spin-off development be contained within urban-area boundaries nearby.[9]

Also, there is a need to examine the scale of developments. While agrotourism developments at wineries remain small, less than 20,000 sq. ft., and relatively dispersed, there are little more than localised negative impacts. But as developments become much larger, their impacts are greater, for example, forcing the hand of urban servicing; and since the wineries are likely not adjacent to existing urban areas, the lands in between may be placed under urban shadow effects, including increasing speculation and raising land prices beyond the means of the agricultural sector. It has been these various pressures which have resulted in St. Catharines and lately the Regional Municipality of Niagara on behalf of all Fruit Belt municipalities devising policies which would strike a good balance between encouraging agrotourism developments but not to the extent that they would jeopardise the primacy of agriculture and the appeal of the area (City of St. Catharines 1999; Regional Municipality of Niagara, 2001).

Smart Growth for Ontario

In addition to agricultural change encouraging preservation of a unique resource, there are several initiatives by the Ontario government at the present time that point in the direction of better leadership and top-down policies and actions concerning a more responsible use of the province's resources. The Niagara Fruit Belt has again been part of the vanguard in this matter as different professional and political lobbying has been recognised at the provincial level.

The concerns about the future of the Niagara Fruit Belt, while constantly on the minds of regional and local planners, have recently become focused by them as a result of the next ten-year review of the Regional Policy Plan. This commenced in early 2000 and was designed to examine what made Niagara a special place in which to live and what policies and strategies were necessary to maintain and enhance that special place (Regional Municipality of Niagara, 2000a). Following public meetings across the region, it became clear that there was an abiding interest in seeing that Niagara retains those qualities that make for a distinctive set of urban and rural communities, and in particular that urban pressures are taken off its unique agricultural lands in order to enhance its wine industry. This led to public interest in looking at Smart Growth as a new approach to development which would improve the quality of life, better protect agricultural and natural areas and make for a more efficient use of municipal resources (Regional Municipality of Niagara, 2000b). A key element here, which harks back to all bygone criticisms of post-war suburban development, is the creation of more compact urban development that builds on existing communities as well as reinvesting more in existing urban areas. It is only in these ways that Niagara's good tender-fruit and grape lands can ultimately be protected. While there has been a legacy of conflicting signals from the Ontario government, there has also been political support at this level for embracing Smart Growth. However, it has to be wondered if the concept is either understood or carries much weight given the strength of the development industry, since the Ontario government at the same time is suggesting enormous freeway expansions in the Greater Toronto Area and considerable urban growth in new suburban communities (The Globe and Mail, 2001).

In Niagara, the debate about Smart Growth remains at the professional and public level; it has not been embraced as yet by regional and local politicians. In the case of the OMB hearing into the expansion of Fonthill, as well as regional and local decisions before this, development concepts and the wider picture could not have been further from any discussion; the hearing and the decision unfurled in the preordained way of consideration of each area or landowner in turn. However, there is a glimmer of hope, relating to decisions taken by the City of St. Catharines, although it must be appreciated that thinking here tends to be ahead of other parts of Niagara. St. Catharines Council is on record, in a unanimous vote, asking the Ontario government to ban further development on Niagara's unique farmland – forever (The Standard, 2000a). The motion was in support of a letter sent by Don Ziraldo, the co-owner of Inniskillin Wines in Niagara-on-the-Lake and a leading member of the industry, to the Premier of Ontario, asking him to preserve Niagara's fruit lands, saying that urban development takes valuable land out of production and makes the remaining land too expensive to buy. While the St. Catharines motion was generally supported in the agricultural community, it was not well received by Regional government, being seen as a criticism of the greenhouse industry being on unique agricultural land and inhibiting the accommodation of future population growth (The Standard, 2000b).

The St. Catharines decision can be said to be revolutionary in two other ways, both of which, could have far reaching effects on urban containment and the preservation of a unique agricultural resource. In asking for a ban on further development, St. Catharines is freezing any future spatial expansion of the city. Once developable land is used up on the western fringe of the urban area, there is no other way to go but further westwards on to unique agricultural land, since Lake Ontario, the Welland Canal and the City of Thorold prevent expansion in the other three directions. While the option of internal redevelopment and intensification is always there, St. Catharines is meanwhile redirecting its traditional suburban development to other municipalities, and in particular to the south. Thinking that St. Catharines is ahead of the pack in preserving a unique agricultural resource should, however, be kept in check. Mounting frustrations concerning the running out of land for traditional urban growth, and the effect this would have on the tax base, recently resulted in city council proposing a study to assess the need to expand the urban-area boundaries to the west (The Standard, 2001c). In effect, the City is tiring of being a trustee of its remaining agricultural land and receiving no financial benefit from the province; and the opportunity to receive such a benefit through an amalgamation or a number of amalgamations of the 13 municipalities in Niagara was shelved by the provincial government after major political backlashes over amalgamations in Toronto, Ottawa and Hamilton. Amalgamation would have allowed St. Catharines to redirect its development to the south and have it remain in the same municipality.

The second issue was the City's request that the Ontario government take over the responsibility for such a unique agricultural resource by creating a permanent agricultural reserve in Niagara and other parts of southern Ontario, in much the same way that has happened in Quebec, British Columbia and California. This move would take decision-making out of the hands of the majority, pro-development politicians in Regional and other local councils by no longer allowing for Official Plan amendments under the Ontario Planning Act and the endless attrition of agricultural land that presently takes place. Any exceptions to the rule would involve a much more difficult and costly procedure at the Cabinet or legislature level in which only a rock-solid case would be worth the effort. Moreover, responsibility transferred to senior levels of government, while not assuring success at every turn and certainly in the process a blow against local democracy, raises the profile and the seriousness of saving a unique agricultural resource, hopefully leading to a marked attitudinal change in all but the most committed devotee of urban development. The Ontario government's response has been muted and falls back on the need for further study. Likewise, the Regional Municipality of Niagara is engaged in an agricultural impact study to assess the importance of the industry to the Niagara economy and new policies that will be needed to support it. In spite of strong protectionist policies on paper, both levels of government continue to vacillate between on the one hand the professional advice, public clamouring and requests from the industry itself to preserve a unique resource, and on the other the traditions of local democracy, the age-old freedoms of landowners to do what they want with their land, and the safety-valve approach of the quick-fix of an economic severance should the going get tough in an agricultural pursuit.

One particular Ontario government study recently completed, while not directly saving one acre of unique agricultural land, does offer the prospect of taking away considerable urban pressure from that land (Ontario Ministry of Transportation, 2001). This relates to the approving of a $1.5 billion, mid-peninsula highway corridor between the QEW in the Fort Erie area and Highway 407 near Hamilton. The major effect of this is to relieve congestion on the existing QEW through Niagara and prevent its expansion from six lanes to ten between Niagara Falls and Hamilton. That alone would have taken even more land out of cultivation in the Fruit Belt; but the improved accessibility that the expansion would give, would have continued to place hundreds of acres of some of Canada's best agricultural lands under considerable threat.

It has been consistently argued ever since the QEW was opened in 1939, that the route through Niagara should have been above the Escarpment on far less sensitive lands; and since the 1950s a mid-peninsula highway corridor has always been a part of provincial transportation planning goals (see, for example, Ontario Ministry of Transportation and Communications, 1976). Until recently, the proposed highway was never seriously considered because traffic demand could always be satisfied by the QEW which was made a limited-access highway in the 1970s and expanded to six lanes as far as Niagara Falls in the 1990s. The 40 per cent increase in truck traffic in the 1990s, resulting from increased trade through NAFTA and just-in-time commercial deliveries, as well as increased tourist traffic, necessitate further highway expansion. The Ontario government's decision reflects more than the relieving of pressures on the Niagara Fruit Belt and the QEW. There is the global need for a mid-peninsula highway corridor to assist the U.S. government in improving communications between the American North-East, Mid-West, South-West and Mexico, including a new border crossing on a more local level, the highway will further encourage the redirection of urban development in Niagara to the southern municipalities and in so doing improve their economic performance and stem some of the criticisms on issues relating to equity between municipalities.

Conclusions

The changing fortunes of the grape and wine industry in Canada, issues such as Smart Growth, agrotourism and transportation, and a more sophisticated public and politician interested in a certain quality of life in Niagara have recently raised the profile of the Niagara Fruit Belt, attracted the attention once again of the Ontario government and re-emphasised the call for the preservation of a unique agricultural resource. However, one would do well not to become overly optimistic that the traditional and expansive nature of urban development will be curtailed, and that government will do any more than study the problem to death and establish restrictive policies that it then circumvents.

For over fifty years now, we have been analysing the problem of expansive urban development and its impact on our agricultural resource base, both in Niagara and elsewhere. The reasons for this development, the effects of it, who is responsible, and the policies that need to be introduced and enforced to manage it have scarcely changed, even though the players have. There continues to be great reluctance among politicians to cut off landowners from having the freedom to do almost what they want with their land, and especially since land and its particular use relate to economic well-being, or lack thereof, and the landowners themselves remain a forceful and vocal lobby. Moreover, since urban development brings greater economic returns than almost all agricultural activities, both the development industry and landowners have always been keen to promote land conversion; and the constant outward expansion of the city means an ever-present threat to our best agricultural lands, unless there are rigorous policies that are enforced. Niagara since 1981 has had these policies, but they have often meant little as regional and local politicians have caved in to one development proposal after another and the Ontario government is hardly to be seen enforcing provincial policy. Niagara's saving grace has fortunately, and ironically, been its slow economic expansion since the 1960s compared to areas closer to Metropolitan Toronto.

Preserving the Niagara Fruit Belt by default is hardly a solution. The pace of economic change has quickened in the last decade, and the status quo of existing urban-area boundaries has been challenged as developable land becomes exhausted. The consequences for the unique agricultural lands in the Fruit Belt could be dire indeed; and the fear continues that the clock will be turned back to the 1950s and eventually all of the Fruit Belt will resemble north St. Catharines or east Hamilton. The way forward to effectively secure the Niagara Fruit Belt as a unique agricultural area is to continue to promote the present developments in the agricultural industry, in particular the grape and wine industry, which naturally has a vested interest in wanting the land base preserved for all time, and conversely would not want to see continuing attrition by urban sprawl. This development, however, must be accompanied by a wholesale shift in attitude on the part of government. The quick-fix through a conversion to urban land use has to be resisted; the already strict regional planning policies on where urban development can take place need to be given real meaning; and since regional and local governments in Niagara are so often to be found wanting, at least at the political level, the Ontario government must play a greater leadership and educational role in both the promotion of an important industry and the preservation of a national resource.

Notes

1 The Niagara Fruit Belt to the west of the Niagara Region boundary, in the recently amalgamated City of Hamilton, has been virtually consumed by residential, commercial and industrial development. From the 1960s onwards, the focus of the debate has shifted almost exclusively to the Regional Municipality of Niagara, and its five Fruit Belt local municipalities – Grimsby, Lincoln, St. Catharines, Niagara-on-the-Lake and Pelham.

2 The Niagara Escarpment, extending through Ontario from Queenston on the Niagara River to Tobermory on Georgian Bay, has a scarp face to the east and north rising up to 200 ft. from the surrounding plain. This significant geological and environmental feature, designated by UNESCO as a world biosphere feature, contains a mix of land uses from natural and wooded areas, to pits and quarries, parklands, agricultural areas, extensive urban areas including Greater Hamilton and St. Catharines, and rural residential estate development. The Ontario government sought to protect this ribbon-like feature by establishing, in 1973, a provincially-appointed Niagara Escarpment Commission (NEC) whose Master Plan took precedence over all overlapping regional and local plans. While regional and local governments have constantly objected to the development restrictions of this undemocratic form of planning administration, and have tried to persuade the Ontario government to abolish it, it can be argued that the NEC is far from being anti-development.

3 Opposition to regional government by Niagara's three cities relates to their loss of powers on becoming lower-tier authorities. In the previous county system of government, in place since the first half of the 19th century, the cities were administratively independent from their counties. On a number of occasions since 1970, there have been pressures brought to bear on the Ontario government to abolish the

Regional Municipality of Niagara and introduce single-tier local authorities with individual (and full) planning responsibilities.

4 PALS's principal witness, who was a full-time academic and the group's researcher when not on the witness stand, was grilled by opposing counsel into a state of complete exhaustion, necessitating an adjournment of the proceedings.

5 The permanency of the urban-area boundaries in the five Fruit Belt municipalities was always disputed at the Regional and local level. It was written into the Regional Policy Plan by the OMB hearing officers as an Appendix, and it was stressed that the 'Region cannot by its Official Plan 'legislate' permanent boundaries between the conflicting activities of the urban and farming communities. It can only provide boundaries that will be recognised and accepted by its inhabitants as being in the best long-term interests of the Region as a whole....and should be regarded as permanent' (Regional Municipality of Niagara 1988, 94). To prevent any further confusion the Ontario government agreed to the removal of the Appendix in 1997.

6 Permission for retirement severances is not granted in all of Ontario's planning jurisdictions. Indeed, a brief by the author, entitled Land-Use Conflict and Sustainable Agriculture on the Rural-Urban Fringe, presented to the Federal government's House of Commons Standing Committee on Agriculture, October 21, 1991, drew surprised comments from rural MPs who maintained that such planning practices, which may be a welcome financial break for farmers, clearly invite unwanted urban influences into an agricultural area.

7 In the late 1980s, the Canadian wine industry feared that it could be wiped out essentially by cheap, Californian imports. The Free-Trade negotiations prevented this from happening, and in return Canada was forced into agreeing to restrict the number of wine retail outlets beyond the wineries themselves and the provincial-government operated liquor stores. The smallness and vulnerability of the Canadian industry can be appreciated when the new vineyards established in South Australia in 1999 (approximately 15,000 acres) were greater than the total acreage in Niagara.

8 While overall the tender-fruit industry may exude a greater air of confidence, there seems a never-ending run of individual farmers, who argue some kind of economic hardship, and seek Official Plan amendments to effect a land-use change and get out of farming. The latest ploy to win favour with Regional politicians, although not with planning staff, is a driving range; it is argued not altogether convincingly that this land use allows for a return to farming (The Standard 2001a, 2001b).

9 Since Niagara sees itself as another Napa Valley, it should be pointed out that since the late 1960s, this Californian wine-producing area has been an agricultural reserve and a ban on most urban land uses has been upheld.

References

City of St. Catharines (1999), *Agritourism,* Draft Report from Planning Services to Staff, File no. 60.2.48.

Commission on Planning and Development Reform in Ontario (1993), *New Planning for Ontario*, Final Report, Government of Ontario, Toronto.

Daniels, T.L. (1991), 'The Purchase of Development Rights: Preserving Agricultural Land and Open Space', *Journal of the American Planning Association*, Vol. 57, pp. 421-431.

Frankena, M.W. and Scheffman, D.T. (1980), *Economic Analysis of Provincial Land Use Policies in Ontario*, Ontario Economic Council Research Studies, University of Toronto Press, Toronto.

Gayler, H.J. (1979), 'Political Attitudes and Urban Expansion in the Niagara Region', *Journal of Urban and Environmental Affairs*, Vol. 11, pp. 43-60.

Gayler, H.J. (1982a), 'The Problems of Adjusting to Slow Growth in the Niagara Region of Ontario', *The Canadian Geographer*, Vol. 26, pp. 165-172.

Gayler, H.J. (1982b), 'Conservation Versus Development in Urban Growth: Conflict on the Rural-Urban Fringe in Ontario', *Town Planning Review*, Vol. 53, pp. 321-341.

Gayler, H.J. (1991), 'The Demise of the Niagara Fruit Belt: Policy Planning and Development Options in the 1990s', in K.B. Beesley (ed.), *Rural and Urban Fringe Studies in Canada*, Geographical Monographs No.21, York University Department of Geography, North York, pp. 283-313.

Gayler, H.J. (1994), 'Urban Development and Planning in Niagara', in H.J. Gayler (ed.), *Niagara's Changing Landscapes*, Carleton University Press, Ottawa, pp. 241-277.

Gayler, H.J. (1996), *The Growing Challenge to Preserving Agricultural Land in Ontario, Canada*, paper presented to the Association of American Geographers Annual Conference, Charlotte, NC.

Gayler, H.J. (2001a), *Fighting Success: Rural-Urban Conflict and the Development of Canada's Greenhouse Industry*, paper presented to the Association of American Geographers Annual Conference, New York City, NY.

Gayler, H.J. (2001b), 'Agritourism Developments in the Rural-Urban Fringe: the Challenges to Land-Use and Policy Planning in the Niagara Region, Ontario', in K.B. Beesley, H. Willward, B. Ilbery and L. Harrington (eds.), *The New Countryside: Critical Questions for the Future of Rural Regions and Communities*, Wilfrid Laurier University Press, Waterloo.

Gertler, L.O. (1968), *Niagara Escarpment Study Fruit Belt Report*, Ontario Department of Treasury and Economics, Toronto.

Government of Ontario (1969), *An Act to Establish the Regional Municipality of Niagara*, Government of Ontario, Toronto.

Government of Ontario (1978), *Food Land Guidelines: a Policy Statement of the Government of Ontario on the Planning for Agriculture*, Government of Ontario, Toronto.

Government of Ontario (1996), *Provincial Policy Statement*, Government of Ontario, Toronto.

Irving, R.M., ed. (1957), *Factors Affecting Land Use in a Selected Area in Southern Ontario*, Ontario Department of Agriculture, Toronto.

Jackson, J.N. (1982), 'The Niagara Fruit Belt: The Ontario Municipal Board Decision of 1981', *The Canadian Geographer*, Vol. 26, pp. 172-176.

Krueger, R.R. (1959), 'Changing Land Use Patterns in the Niagara Fruit Belt', *Transactions of the Royal Canadian Institute*, Vol. 32(2), pp. 39-140.

Krueger, R.R. (1982), 'The Struggle to Preserve Specialty Crop Land in the Rural-Urban Fringe of the Niagara Peninsula', *Environments*, Vol. 14, pp. 1-10.

Krushelnicki, B. (1994), 'The Progress of Local Democracy in Niagara: The Evolution of Regional Government', in H.J. Gayler (ed.), *Niagara's Changing Landscapes*, Carleton University Press, Ottawa, pp. 325-349.

Miller O'Dell Planning Associates (1997), *Background Report to an Application for a Zoning By-Law Amendment regarding the Seventh Street Market*, St. Catharines.

Niagara Region Local Government Review Commission (1966), *Report of the Commission*, Department of Municipal Affairs, Toronto.

Ontario Ministry of Transportation (2001), *Niagara Peninsula Transportation Needs Assessment Study*, Toronto.

Ontario Ministry of Transportation and Communications (1976), *Niagara-Lake Erie Transportation Study*, Toronto.

Ontario Municipal Board (2000), Decision relating to Official Plan Amendments to allow urban expansion in Fonthill, File no. PL980963.

Ontario Tender Fruit Producers' Marketing Board (1999), *Production and Market Opportunities in the Ontario Tender Fruit Industry*, St. Catharines.

Reeds, L. (1969), *Niagara Region Agricultural Research Report*, Ontario Department of Treasury and Economics, Toronto.

Regional Municipality of Niagara (1988), *Regional Niagara Policy Plan*, Office Consolidation.

Regional Municipality of Niagara (1991), *Amendment 60 (Revised Agricultural Policies).*

Regional Municipality of Niagara (1992), Autotrend Transportation Centre, File: RE.am.52.

Regional Municipality of Niagara (1995a), *Niagara Tender Fruit Lands Program: Proposed Administrative Details*, report by the Agricultural Easement Committee to the Minister of Agriculture, Food and Rural Affairs.

Regional Municipality of Niagara (1995b), *Modified Land Use Regulations to Support the Niagara Agricultural Industry*, PD 114-95.

Regional Municipality of Niagara (1997), *Regional Policy Plan Amendment 111: Revised Agricultural Policies*, DPD 110-97 (Revised).

Regional Municipality of Niagara (1998), *Regional Policy Plan Amendment 120: New World Culinary Centre, Town of Lincoln*, DPD 47-98.

Regional Municipality of Niagara (1999a), *Regional Policy Plan Amendment 138: Port Robinson West Urban Area, City of Thorold*, DPD 144-99.

Regional Municipality of Niagara (1999b), *Proposed Expansion of the Fonthill Urban Area, Town of Pelham*, DPD 136-99.

Regional Municipality of Niagara (2000a), *Niagara – A Special Place Today and...Tomorrow?* Publication No. 95-1.

Regional Municipality of Niagara (2000b), *Smart Growth for Niagara: A New Approach to Development*, DPD 118-2000.

Regional Municipality of Niagara (2001), *Comparison of Agri-tourism Policies: Town of Lincoln, Town of Niagara-on-the-Lake and City of St. Catharines*, DPD 10-2001.

The Globe and Mail (2001), 'Urban sprawl threatens prime farmland', Toronto, June 11, A15.

The Standard (2000a), 'Council Urges Ban on Farmland Development', St. Catharines, July 27, A3.

The Standard (2000b), 'Farmland Protection Motion Criticized', St. Catharines, August 10, A6.

The Standard (2001a), 'Fruit lands' new threat – golf balls', St. Catharines, April 12, A3.

The Standard (2001b), 'Why allow driving ranges on our tender-fruit lands?', St. Catharines, April 14, A12.

The Standard (2001c), 'City Tired of Guarding Farms', St. Catharines, July 10, A1.

Chapter 5

Planning and Land Regulation at Lake Tahoe: Five Decades of Experience

Robert H. Twiss

Introduction

The beauty of Lake Tahoe was captured eloquently by Mark Twain some 130 years ago:

> ...a noble sheet of blue water lifted six thousand three hundred feet above the level of the sea, and walled in by a rim of snow-clad mountain peaks that tower aloft a full three thousand feet higher still!...As it lay there with the shadows of the mountains brilliantly photographed upon its still surface I thought it must be the fairest picture the whole earth affords. (Twain 1872)

Today, in spite of environmental decline, Lake Tahoe remains one of the clearest large lakes in the world, and the lake and its watershed are still worthy of the best stewardship we can provide. Indeed, in July of 1997 President Clinton and Vice President Gore convened a 'Lake Tahoe Presidential Forum', with the President referring to Lake Tahoe as '... one of the crown jewels, unique among them all it's a national treasure that must be protected and preserved'. (Whaley, 1997) However, the ecosystem that supports the lake and watershed's scenic beauty is under threat from the impacts of urban growth. Federal Judge Edward C. Reed, in deciding a recent property rights case, wrote:

> The lake is almost indescribably beautiful. The Nevada Supreme Court called the lake a 'national treasure' (Kelly, 1993), and the California Supreme Court referred to the Basin as 'an area of unique and unsurpassed beauty' *(People v. Younger, 1971).* Ironically, the more Lake Tahoe comes to be appreciated for its beauty, the more that beauty is threatened (TSPC, 1999).

While Judge Reed's fact-filled 72 page opinion demonstrated his deep appreciation of the region's value and his mastery of the complex science underlying ecosystem management; he nonetheless felt compelled to decide that the Tahoe Regional Planning Agency's land protection regulations constituted a taking of private property (TSPC, 1999). This decision was reversed by the Ninth Circuit Court of Appeals, and the validity of the regulations was finally affirmed by the U.S. Supreme Court (United States Supreme Court, 2002). Given the breadth and depth of the Supreme Court's recent opinion, the planning and regulatory structure at Tahoe have taken on added significance.

Purpose

This chapter presents an overview of regional planning and land regulation issues in the Lake Tahoe Basin. It sketches the history of land development over the past fifty years, and the evolution of the planning structure of the Tahoe Regional Planning Agency (TRPA).

The Tahoe experience offers many lessons for planners working in complex environments. It was a prototype for eco-region based planning. Many of Tahoe's problems stem from a mismatch between the geography of governmental jurisdictions and the lake's physical eco-region, which spans five counties in two states. This lack of congruence has been fixed, and the entire lake and watershed are now under the jurisdiction of one agency. The TRPA has full authority to set environmental standards, to plan, and to grant or deny all manner of land development permits. Thus Tahoe offers lessons in both the evils of mismatch, and the benefits of congruence.

Lake Tahoe also presents a case study for 'mop-up' planning. At Tahoe, as in many other places, no comprehensive plan was adopted until after the region's environmental capacities were seriously over-committed. Thus, the main thrust of planning and regulation has not been one of forward-looking, clean-slate visioning, but rather one of trying to pick up the pieces and undo the damage caused by flawed decisions. Mop-up planning is characteristic of virtually all of the world's ecosystem restoration efforts.

Planning for Lake Tahoe is also at the center of important property rights conflicts. All the forces above combine to put Tahoe at the cutting edge of the property rights debate. Thousands of subdivided single-family lots were permitted in stream zones and wetlands, and on steep slopes. In spite of government buy-out programs, many landowners are seeking judicial relief.

The complexities of the former issues exacerbate the high risk of planning for ultra-fragile environments. Lake Tahoe's alpine setting, and the chemistry and biology of the lake itself leave little room for experimentation or error.

Within this framework, ultimately a capacity-based planning model was developed and implemented. While early land development decisions were ad hoc, the current system is based on two pillars:

- Land Capability System. Under the first Interstate Compact (1969), a science-based system ranked all lands in the Tahoe Basin as to their relative fragility, and restricted land disturbance and coverage accordingly. While not perfect, and while deployed much too late in the game, this system has withstood thirty years of scrutiny and legal challenges and still supports the basic structure of today's planning and regulation.

- Environmental Threshold Carrying Capacities. Under the second Interstate Compact (1980) firm levels of protection for the lake were established, and a new plan created to meet these standards. The concept of 'Standard-driven planning' is new, and Tahoe provides perhaps the earliest clear example of this approach. Taken together, the Land Capability System and Threshold Carrying Capacity templates provide what may be the most fully developed ecosystem-based plan in the world.

Early History of Land Development and Environmental Concerns

The Lake Tahoe Basin lies about 150 miles northeast of San Francisco; about a five-hour drive from the Bay Area and only about two-hours from the burgeoning Sacramento Valley. Both the lake itself and its watershed are split by the California/Nevada border, and then further divided into four counties and two cities. In the early years of development, none of the seats of local government were in the Tahoe Basin itself. Responsibility and authority for stewardship were fragmented, and no one was minding the store (Figure 5.1).

As late as the 1940's, the region was very lightly populated with a few summer resorts, large estates and cottages along the lakeshore. About half the land was in National Forest and most of the rest in large private holdings for timber and summer grazing of cattle. This was the era of slow, incremental subdivision of large parcels into smaller lots. There was no zoning as such, with the area in 'unclassified' county zoning status.

In the mid 1950's pressures were intensified by the upgrade of US Highway 50, providing year-round access, and the development of Heavenly Valley ski area at South Lake Tahoe. As of 1955 there were about 20,000 individual parcels of land throughout the entire 300 square mile watershed, but in 1955 alone, over 1000 new subdivided parcels were created.

At this juncture, far-sighted individuals began to express concern about the pace and form of development, and in 1957 the Max Fleishman Foundation helped establish the first basin-wide group to develop an informal but comprehensive land-use plan. Federal funds to support formal planning were not forthcoming, unfortunately, but development went forward without planning, with more than 2000 new vacant lots created that year. These 'lot-and-block' type subdivisions were required to put in streets and water, but this was before the era of Planned Unit Development and the simultaneous construction of houses. Subdivision created thousands of vacant lots and thousands of individuals with the expectation of building someday. The problem with lot-sales programs is that they pass land-use problems on to individual buyers who are not well equipped to anticipate and solve complex problems of on-site development, or to control or mitigate off-site impacts.

A bi-state multi-county planning group was formed in 1958, but while this group studied and shared information, it had no authority, and a short life (Jackson and Pisani, 1973). In 1959, dredging and filling of the marsh at the Upper Truckee

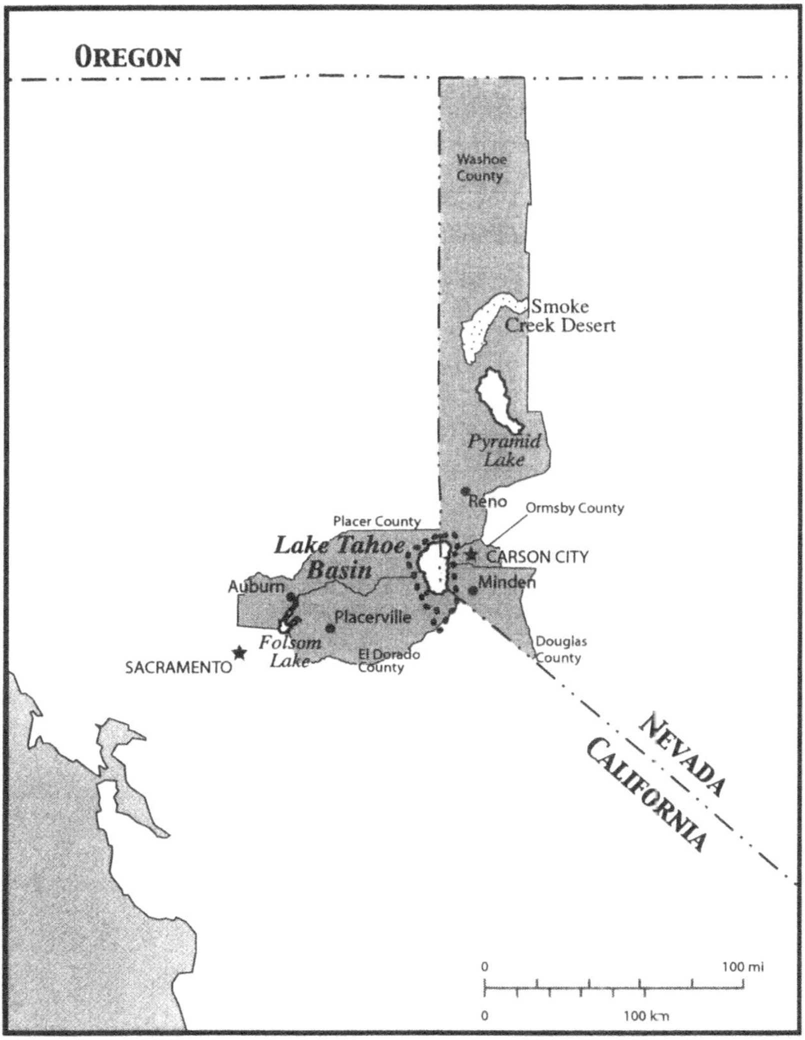

Figure 5.1 Lake Tahoe Region

River's inlet to Lake Tahoe for the Tahoe Keys housing subdivision was begun, and the basin-wide about 2500 new residential lots were created. By 1960, lake pollution was becoming apparent, and the Nevada State Engineer issued a strong warning. The Winter Olympic games were held at Squaw Valley just outside the Basin, heightening development pressures; and 2800 additional lots were approved.

In the period 1960-62, there were more studies of pollution problems, but no action, and the vast Incline Village subdivision was approved even though a solution to sewage problems was not at hand. A draft basin-wide plan combining the desires of the local governments would have accommodated a sixteen-fold increase in population to 418,000.

By 1965 this draft plan had been revised downward, but still would have resulted in a population of 313,000. For the first time, the legislatures of the two states convened panels to discuss the bi-state Lake Tahoe Basin. Each of the five separate counties deliberated the draft plan, with one county adopting, but others remaining undecided. Nonetheless, the individual local governments created some 3,000 more lots in 1965.

Finally, in 1967, the Nevada and California legislatures held the first formal public hearings on the concept of a bi-state agency. California created its own separate agency to oversee the California side of the lake, and that same year the pollution problem literally came to the surface with a 2,000,000 gallon-per-day sewage overflow on the north shore of the lake.

The intense political activity to curb growth did not go unnoticed by the real estate and development community, and created a rush to get vested with the necessary permits before restrictions could be put in place. Land developers sought to gain 'in-concept' approval of multi-phase subdivisions. Small lot owners rushed to excavate and get foundations in place so as to gain an entitlement to build later. Local governments acting independently, and without a comprehensive regional plan, approved subdivisions according to whatever zoning laws were in place. So while only 913 lots had been created in 1967, the number increased to 1,638 the next year, doubled again to 3,462 the next year, and by 1969 the year that Congress finally ratified the Interstate Compact creating TRPA, approvals reached their peak of 4,134 new parcels in one year.

Environmental Impacts of Land Development

The urban development rush of the late 1960's caused several different types of environmental impacts. Problems stemmed from actions occurring at a number of critical points in ecological cycles. Planners at Tahoe realized early that any useful plan and supporting regulations would have to take account of all of the key types of impacts, and address their true causes (Twiss, 1994).The principal impacts were identified early and remain operationally critical today (TRPA, 1982). They include:

- Soil erosion. De-vegetation and grading and cause accelerated erosion and subsequent movement of soil particles to the lake. Construction can cause problems, even though the site may be far from Lake Tahoe. While individual houses and driveways create relatively small problems, the cumulative impacts are significant.
- Transfer of soil and nutrients from the site of disturbance to streams and the lake. Building in steep terrain or on poor soils accelerates soil erosion. Streets serve as conduits, moving water and silt from the development site to the nearest stream and thence to the Lake.
- Interception of ground water. Grading and land leveling can intercept the near-surface ground water table. Water allowed to remain in the ground will have nutrients taken up by vegetation and released to the air. But exposure of the ground water table can cause nutrient-rich water to emerge at the surface where nutrients can be easily transmitted to streams and the lake.
- Alteration of stream channels. Stream channels and streamside vegetation can help slow or reduce the throughput of nutrients to the lake. However, if disturbed, streams can lose these functions and add problems through stream-bank erosion.
- Destruction of wetlands. Early development included dredging and filling of wetlands, including major destruction of the marsh at the outlet of the Upper Truckee River by the Tahoe Keys subdivision.

The cumulative effect of the impacts has been to hasten the aging process of the lake. The buildup of nutrients facilitated the growth of algae, which in turn reduced the transparency of the lake's waters. Reduced transparency concentrated the sun's energy in the upper layers of the lake's waters, reducing the degree of mixing, and further increasing the potential for algal growth. (Goldman, 1994)

The First Interstate Compact and the Creation of TRPA

While environmental impacts continued virtually unabated, the planning process for the Tahoe region consistently lagged behind the development boom. Indeed, it took some twelve years to move from the first study recommending basin-wide planning to the point where President Nixon signed the Interstate Compact. The process of creating the interstate compact was time consuming and arduous. Both state legislatures had first to pass identical legislation and both Governors had to sign as well. Since the Nevada legislature met only every two years, this slowed the process greatly. The process was so lengthy, and the pre-plan land rush so great, that it can be argued that Tahoe may never be able to recover from this unplanned development. In spite of today's aggressive planning and regulation, and the millions of private and public dollars being spent in land purchases and ecosystem restoration, it's not known if we will be able to save Lake Tahoe.

The main strengths of Tahoe Regional Planning Authority under the first compact were numerous. For the first time, the lake and its entire watershed came under one jurisdiction. Even today, thirty years later, there are few such examples of multi-jurisdiction planning worldwide. Secondly, the new agency was given broad authority to regulate all aspects of land use. This included the setting of air and water standards normally reserved to the federal government and states, to the designation of the type, intensity and location of land use (normally the prerogative of local governments) Finally, the first TRPA plan and ordinances adopted in 1972 included a new 'Land Capability System' to determine the intensity of land use to be permitted. New development of all types would be allocated to low-hazard lands (Compact, 1969). As with all policy frameworks, there were structural weaknesses that were evidenced during implementation. They included the following six issues. First, during the planning period, there was no development moratorium. The newly created agency spent roughly half its time approving still further land subdivision under weak county zoning as the planning process was undertaken. Consequently, before the first TRPA plan and ordinances could be adopted in 1972, TRPA itself had approved 4,600 additional lots. Secondly, subsequent to the passage of the 1972 TRPA Plan, urbanization and environmental impacts continued. It was not so much that TRPA failed to carry out its Congressional mandate, but rather that the original compact and plans were not equal to the task.

Third, while the compact conveyed sweeping powers, at the same time it curbed the use of that power with a weak-voting structure. Under the first compact, an affirmative vote of a majority of members from each state was required to deny a project. Since the two states had very different agendas, stalemates were common; and many projects meeting the minimum standards of the new plan and zoning were approved by default. A fourth problem was grandfathering, more than 20,000 previously created subdivided lots required only building permits to proceed with construction. Nearly half of these were in high-hazard areas, but they were allowed under the first Compact. Fifthly, the 1969 Compact's guidance to the agency's planning was to strike a balance between the region's economy and environmental protection, which in fact led to a continued degradation of the environment. Finally, the 1969 Compact made the finding that 'the environment of the Basin may be threatened by development'. This language forced the agency to prove and reprove the impacts of every separate action.

Planning and Regulation Under the First Interstate Compact

The planning and regulatory structure was crafted to respond directly to the key types of impacts described above. They addressed an extremely wide range of situations, and targeted them as precisely as possible. Careful targeting was necessary to avoid unwarranted regulation of properties situated in robust environmental situations, yet fully protect critical problem areas. During the first planning period, 1968 to 1971, planners confronted several difficult issues to spatial and environmental attributes of all of the Tahoe region.

The ratio of the lake to watershed presented special problems in land planning and regulation. Roughly 200 square miles of the watershed's 500 square miles of area is taken up by the lake itself. The watershed is small for a lake the size of Tahoe, and small changes in the watershed can have commensurately large impacts. Another ramification of the lake/watershed ratio is that the lake's volume represents about 700 years' precipitation, and this presents another aspect of the region's fragility. With a residence time of about 90 years, nutrification and pollution must be considered as irreversible.

Another special issue was the basin's extremely wide range of environmental conditions. Thus making the planning and regulatory job especially difficult. The region includes true alpine conditions above 9,000 feet elevation, with solid granite rock, no timber, no groundwater, and severe climatic stress. But it also includes environments at lower elevations around 6300 feet, for example, the depositional lands formed by the flood plain of the Upper Truckee River: nearly flat, with deep soil, near-surface groundwater, wetland vegetation, timber, and relatively minor climatic stress. In between, are slopes of widely varying composition and impact potential. The underlying geology ranges from unweathered granite, to unstable glacial moraines and stream-cut volcanic deposit.

Consequently, the TRPA's planning and regulatory structure has had to address all points in the hydrologic/ecologic cycle: uphill erosion, hydrologic conveyance, stream-zone protection, and wetlands protection. A strong scientific base was established at the outset of TRPA and Forest Service planning, and knowledge about the region's particular environmental conditions, and cause-and-effect environmental relationships have continued to be applied as the basis of plans and ordinances.

Early environmental planning at USFS/TRPA was directly based on an assessment of natural factors. The Land Capability System was first developed in 1968 by the U.S. Forest Service planning team for the Tahoe Basin, and enhanced as it was implemented in the early 1970s. (Bailey, 1974) In practical terms, this system is similar to the approach best described by Ian McHarg (1969) but which was developed earlier and then in parallel with McHarg's work. Some sixty environmental factors were evaluated and mapped in one of the first applications of geographic information systems (GIS). From this assemblage, a few layers were selected as being most related to land disturbance, including elevation, slope, aspect, soil type, geomorphic type, vegetation class, and rainfall intensity.

Even before these planning efforts were begun, the lake's ecology was being studied by University of California Davis limnologist, Charles R. Goldman (Goldman, et al, 1993). His long-term ecological research and monitoring efforts had been tracking the lake's steady decline in clarity, and its increasing nutrification. Goldman gave early warnings of lake's decline, and correctly identified the cause-and-effect relationship between land development and the premature aging of lake. Of paramount concern were nitrogen and phosphorus that supply the nutrients for algal growth. These elements come from sewage, urban and landscape runoff, and soil erosion. The sewage-treatment problem was to be solved over the long run by basin-wide collection, treatment, and export. The remaining serious problem addressed by the land planning and regulatory effort was that of pollution due to land disturbance and development.

Land development has three key types of impacts. First, the removal of natural vegetation that leads to accelerated soil erosion and loss of nutrient uptake on the other. Secondly, trenching and grading leads to concentration and acceleration of runoff. Finally, the creation of impervious surface from roads, parking areas, driveways, and roofs which intercepts precipitation and delivers it to culverts and streams; and in the process, bypasses natural processes of nutrient uptake.

In devising the Land Capability System, these types of physical actions were simplified and combined into a standard impact. Effectively, this generic standard imprint was placed over maps of key environmental factor, including slope, aspect, elevation, soil type, surficial geology, and natural vegetation. A multi-disciplinary team of experts, composed of a forester, soil scientist, geomorphologist, civil engineer, and landscape architect, using professional judgement and field verification, ranked the relative ability of each mapped area to withstand and recover from imposition of the standard imprint. Using soil-type boundaries as a guide, a single composite expression of this ranking was assembled into the Land Capability Map. The map classifies the watershed into thirteen types of situations, representing seven levels of capability. These varied from Class I (Steep slopes, high elevation, wetlands, stream zones, and other areas unable to withstand any significant disturbance), to Class VII (gentle slopes, deep soil, lower elevation, well vegetated) able to absorb and recover from substantial alteration.

Many regional plans have used a similar approach to environmental analysis, but what made the first Tahoe plan was that the Land Capability Map was adopted by ordinance, and implemented through limits on the amount of impervious surface. Its legal stature is equal to that of a zoning type map that governs the type and intensity of land use in terms of dwelling units and commercial square-footage. For example, a parcel in Land Capability Class I is to be limited to 1 per cent land coverage even though it might be zoned Commercial. Conversely, another Commercial parcel in Class VII could cover up to 30 per cent with impervious surfaces. This 'Two-map' planning system is currently in force.

Planning Under the Second Interstate Compact and Subsequent Litigation

Almost immediately upon enactment of the first Interstate Compact, work began to remedy its main weaknesses. However, little could be done without first letting the TRPA planning process operate its course for several years to see if in fact the perceived weaknesses were real. In 1974, a U.S. EPA study called for strengthening, of Tahoe's planning efforts but the two state legislatures could not agree upon changes. Bills were considered again two years later, but again without agreement. In Congress, proposals were considered to declare the region a National Recreation Area if the Compact were not strengthened. Finally, in 1980, under threat of federal action, a revised compact was adopted by both states, and quickly ratified at the federal level (Compact, 1980).

The 1980 Compact made an explicit finding that the region's environment was indeed imperiled, and that land development activities required control and mitigation. This finding greatly simplified the project review and EIS process. The voting structure of the TRPA Governing Board was changed to remove the presumption of approval, and made it more difficult for the Board to amend the regional plan. The most important planning change was to require TRPA to set standards that would ensure the protection of both the lake and watershed, and to make sure that the regional plan would not violate standards. The compact required that 'Environmental Threshold Carrying Capacities' be established. Further, an EIS on the subsequent regional plan had to specifically ensure that the rate, location, and intensity of land development would not jeopardize attainment of the established thresholds, and that the governing Board was required to make findings to that effect in the approval of each project.

As was the case in the National Environmental Policy Act, and state legislatures Congress were mandatory a process that was not yet operational for planners and the scientific community. The concept of carrying capacity has a long history, and is a common term in biology and conservation texts. However, application to real-world problems has been limited, and mainly limited to forest and range management. Nonetheless, TRPA did carry out its Congressional mandate.

The Tahoe Environmental Threshold Carrying Capacities that were implemented and adopted are notable in several respects. (TRPA, 1982). First, they are unusually comprehensive, establishing clear standards for environmental characteristics that are normally proscribed by more than a dozen separate agencies at local, state and federal levels. In addition, the thresholds set numerical standards for subjects that are important but difficult to measure, such as scenic resources, and that elsewhere are covered only by verbal policies, if they are considered at all. The major categories of thresholds are:

- Water Quality. Standards cover Pelagic Lake Tahoe (phytoplankton and primary productivity), and Secchi disk transparency (the basic measure of clarity). They also cover the littoral zone of Lake Tahoe (e.g. the near-shore waters), with standards for dissolved inorganic nitrogen, turbidity, dissolved phosphorus, iron, other nutrients. Separate thresholds cover the tributaries to the lake, with standards for elements above, plus suspended sediment. There are standards for surface runoff (inorganic nitrogen), and for groundwater.
- Soil Conservation. The thresholds carry forward the land-coverage standards embodied in the previous Land Capability system.
- Stream Environments. Thresholds protect such zones from further encroachment, but go further to establish a requirement that a substantial portion of previously damaged stream zones be restored.
- Air Quality. Thresholds cover the most commonly regulated forms of air pollution (e.g. Carbon Monoxide, Ozone). But they go further, and regulate items of special significance to the Tahoe situation: regional visibility, particulate concentrations, wood smoke as a per cent of 1981 base value, nitrate

deposition relative to lake nutrification, and vehicle miles traveled to achieve a 10 per cent improvement over 1981.
- Other thresholds cover vegetation, wildlife, fisheries and lake habitat, noise, and scenic resources.

Most importantly, the thresholds are much more that just a list of goals. All land development projects bear the burden of proof in showing that they will not slow progress toward attainment.

Given the scope of coverage required by the compact and the difficulty in setting numerical standards for poorly defined values, it is not surprising that it took TRPA about two years to develop the thresholds. And, it took another year and a half to create a plan that would implement them. Thus, it was not until May of 1984 that the revised plan and EIS were adopted. This planning package was deemed too weak by the League to Save Lake Tahoe and the California Attorney General and deemed too strong by the property-rights group: Tahoe Sierra Preservation Council, all of whom immediately filed suit. This led quickly to a Federal Court Preliminary Injunction blocking TRPA from implementation. Virtually all private-land development projects were held in abeyance for three years while revised plans and regulations were developed.

The revised plan made a number of changes in land intensity, but its most significant addition was a re-casting of the Land Capability System in its application to small vacant parcels. Based upon a re-mapping of environmental conditions and two-year's fieldwork, each of some 14,000 vacant lots was subject to on-the-ground site inspection and ranking as to its development suitability. Individual Parcel Evaluation System (IPES) scores were used to determine which lots could be built upon, and also guided public acquisition programs.

Current Status and Conclusions

Since 1987 there have been relatively few changes to the planning structure and in-fill development has gradually filled out the plan. Given the tortuous history of this planning and regulatory structure, it is not surprising that it is highly resistant to modification. Newer concepts, such as adaptive ecosystem management, or transfer of developmental rights (TDR) tried. For better or worse, the regulatory system is relatively brittle, and is flexible only through actions by the courts.

The general effect of the thresholds and the embodied Land Capability System has been to freeze the historic pattern of land use, and to dramatically limit new development to pre-existing lots and commercial parcels. This is not because the old pattern is desirable, nor that it would not be desirable to re-cast and re-develop, but because the room to maneuver was lost during the land-rush era. Re-shaping the pattern to better fit land capabilities and transit objectives through TDR or similar tools is nearly out of the question because of the impacts that would come from further raw-land disturbance. The urban boundary of Tahoe City has been virtually walled-in by mapped stream environment zones, and it has been almost impossible to re-shape the initial pattern of land use, to undo strip-commercial development, or to improve parking and pedestrian circulation. For example, the South Lake Tahoe airport no longer has commercial air service, in part due to economics and difficulties in meeting the navigation and safety requirements posed by Tahoe's terrain, but also in part to the threshold's strict noise standards which preclude operation of large air craft.

There have been isolated but important examples of urban redevelopment in the city of South Lake Tahoe, which adjoins Nevada's casino/hotel zone. However, this very successful effort has been costly, and not replicated throughout the Basin.

The California and Nevada land conservancies and the U.S. Forest Service have aggressive and well-funded programs that have purchased low capability lands, including both large parcels and small vacant lots. Whereas thirty years ago the Basin was split about evenly between public and private land, now only about 20 per cent remains in private ownership.

While much can be learned from the history of development and planning at Tahoe, there is much still to be watched. The fate of the lake is still in the balance, and we have yet to see if the measures taken thus far, stringent as they are, will be successful. The scientific and technical underpinnings of the plan and ordinances have stood up well, and may serve as models for aspects of work elsewhere.

It is not by chance that in 2002 the Supreme Court chose a Tahoe case to re-think and restate its principles in regard to land-use planning private property. In this case, the Court was presented with a rich set of decision records and trial documents, and the implications of legal principles could be clearly seen. The Court chose to review its earliest land-use cases, and restate its intentions in cases over the past two decades. While they might have limited their findings to the issue of a planning moratorium, they expanded their opinion to reaffirm those takings doctrine that should be considered in relation to the parcel as a whole, not just one strand in the bundle of rights. And they reaffirmed the need for judges to weight an array of factors, following the guidelines laid down in *Penn Central* (U.S. Supreme Court, 1978).

The reach and precision of the planning tools crafted at Tahoe will undergo continued scrutiny, especially as they impact property rights. But it appears that in regard to the scientific basis of planning, and in the application of planning and regulation, the TRPA structure has stood up well under fire.

References

Compact (1969), *Tahoe Regional Planning Compact*, Pub. L. No. 91-148, 83 Stat. 360.

Compact (1980), *Tahoe Regional Planning Compact*, Pub. L. No. 96-551, 94 Stat. 3233.

Goldman, Charles R. (1994), 'Anthropogenic Impacts on Oligotrophic Lake Systems Emphasizing Lakes Tahoe and Baikal,' in H. Sund and Y. Xiaogan (eds.), *Environmental Protection and Lake Ecosystems,* Chinese Science and Technology Press, Beijing.

Goldman, Charles R., Jassby, Alan D. and Hackley, Scott H. (1993), 'Decadal, Interannual and Seasonal Variability in Enrichment Bioassays at Lake Tahoe, California Nevada, U.S.A.' *Canadian Journal of Fisheries and Aquatic Sciences*, Vol. 50(7-8), 1489, 1491.

Jackson, W. Turrentine and Pisani, Donald J. (1973), 'From Resort Area to Urban Recreation Center: Themes in The Development of Lake Tahoe, 1946-1956', Environmental Quality Series No. 15, University of California, Davis, CA.

Kelly vs. TRPA. (1993), 855 P 2d. 1027, 1034 (Nev. 1993).

Las Vegas Review Journal (1997), 'Clinton Pledges Tahoe Aid,' July 27, p. 1A.

McHarg, Ian L. (1969), *Design with Nature,* The Natural History Press, Garden City, NY.

People ex rel. Younger vs. County of El Dorado (1971), 487 {.2d 1193, 1194 (Cal. 1971).

Tahoe Regional Planning Agency (1982), *Study Report for the Establishment of Environmental Threshold Carrying Capacities*, Zephyr Cove, Nevada.

Tahoe-Sierra Preservation Council (1999), *Tahoe-Sierra Preservation Council vs. TRPA*, CV-N-84-257-ECR, US District Court, District of Nevada.

Twain, Mark (1872), *Roughing It*, American Publishing Company, Hartford.

Twiss, Robert (1994), http://www.regis.berkeley.edu/tahoe/tahoegen.html.

United States Supreme Court (1978), *Penn Central vs. New York City*, 438 U.S. 104.

United States Supreme Court (2002), *Tahoe-Sierra Preservation Council Inc. vs. Tahoe Regional Planning Agency*, No. 00-1167.

USDA, Forest Service (1974), Bailey, Robert G., *Land Capability Classification of the Lake Tahoe Basin, California Nevada: A Guide for Planning*, ER 11.

Chapter 6

The Everglades: Where Will All the Water Go?

Jaap Vos

Introduction

South Florida is home to over six million people, it is also home to a unique ecosystem, the Everglades. The relationship between the Everglades and South Florida's residents has been a stressful one. Initially, the people of South Florida perceived the Everglades as a useless, mosquito infested swamp that needed to be drained and made suitable for agriculture and development. Then, they viewed it as an area from which they needed to be protected with levies, canals and an intricate pumping system. Recently, the people of South Florida have realized that the Everglades provides for their water and is a major reason for a thriving tourist industry. It has also become clear that human activities have altered the Everglades system dramatically and that the system and thus the well-being of South Florida's residents, is in serious peril. With this, a new chapter in the relationship between South Florida's people and the Everglades has begun. A chapter in which the natural Everglades and the people of South Florida will have to find a way to exist together in mutual harmony.

This chapter describes the sometimes painful relationship that has existed between the people of South Florida and its crown jewel, the Everglades. The chapter starts with a description of the original (pre-drainage) Everglades. During this time, the Everglades were home to approximately 10,000 Calusa, Jeaga, Jobe and Tequesta Indians that lived in relative harmony with their surroundings. These original South Florida Indians had adapted a lifestyle that allowed them to take full advantage of the opportunities that the Everglades offered them without any major impacts on the ecosystem (McCally, 1999). This chapter then discusses the engineered Everglades with its intricate system of levees and canals, and the events that led to the engineered system. The chapter continues with a discussion of the problems with the existing system and a description of the Everglades Restoration Plan. Finally, the consequences of the restoration for urban development, population growth and social equity in South Florida are discussed.

Pre-Drainage Everglades

The Florida Everglades were part of the larger Kissimmee-Okeechobee-Everglades watershed which stretched over half of the Florida peninsula from the Kissimmee River, just south of Orlando, to Florida Bay (Figure 6.1). The entire watershed spanned 250 miles from North to South, 60 miles East to West and it encompassed an area of about 9000 square miles (McPherson and Halley, 1996). The area known as the Everglades consisted of the southern half of this watershed, stretching about 100 miles from the southern tip of Lake Okeechobee to Florida Bay and covering about 4,500 square miles.

Within the Kissimmee-Okeechobee-Everglades watershed, water from as far North as Orlando flowed through the meandering 107 miles of the Kissimmee River into Lake Okeechobee. Consisting of 470,000 acres, Lake Okeechobee is the second largest freshwater lake entirely within the United States. However, it is extremely shallow with an average depth of just 12 feet. Consequently, Lake Okeechobee would periodically overflow and the water would continue south as a wide thin sheet of water, the famous River of Grass, until it slowly found its way to Florida Bay and the Gulf of Mexico (U.S. Army Corps of Engineers and South Florida Water Management District, 2000). The River of Grass covered most of south Florida and was only bound by the Atlantic Coastal Ridge on the east and an inland ridge on the west. During the rainy season, from May to October, the River of Grass could reach depths of three to four feet at its deepest point while flowing at a speed of about 100 feet per day. In the dry season, the River of Grass would almost completely dry up leaving a landscape dotted with small pools of water.

The Everglades provided refuge for a large number of wading birds such as Wood Storks, Ibises, Herons and Egrets, that nested in mangrove estuary of Florida Bay, Corkscrew Swamp in Southwest Florida and along the Kissimmee River. Large predators such as crocodiles, alligators, panthers, and black bears were also common. Fish, such as snook, tarpon, sea trout, and red drum, were abundant in the estuaries.

Everglades plants and animals were adapted to alternating wet and dry seasons. When water levels gradually dropped during the dry season, fresh water fish would be forced to migrate to deeper pools. Birds, alligators, and other predators would then concentrate around the pools and feed on the large amount of fish concentrated in the pools. During the rainy season, wildlife would disperse throughout the entire watershed and repopulate the Everglades.

The system was rejuvenated by fires caused by lightning strikes during storms in the summer's wetseason. Fires would burn through grass and forest understories until they hit a natural barrier such as water or mangrove, or until they were extinguished by heavy rains. They would leave a mosaic of burned and unburned vegetation in their path which provided the diversity of habitats needed for native plants and wildlife.

When the Spaniards arrived in South Florida in the early sixteenth century, the area was occupied by approximately 10,000 to 30,000 Calusa, Jeaga, Jobe and Tequesta Indians (McCally, 1999, p. 42). Most of the Indian settlements were at the southwest coast, with additional settlements around Lake Okeechobee and the

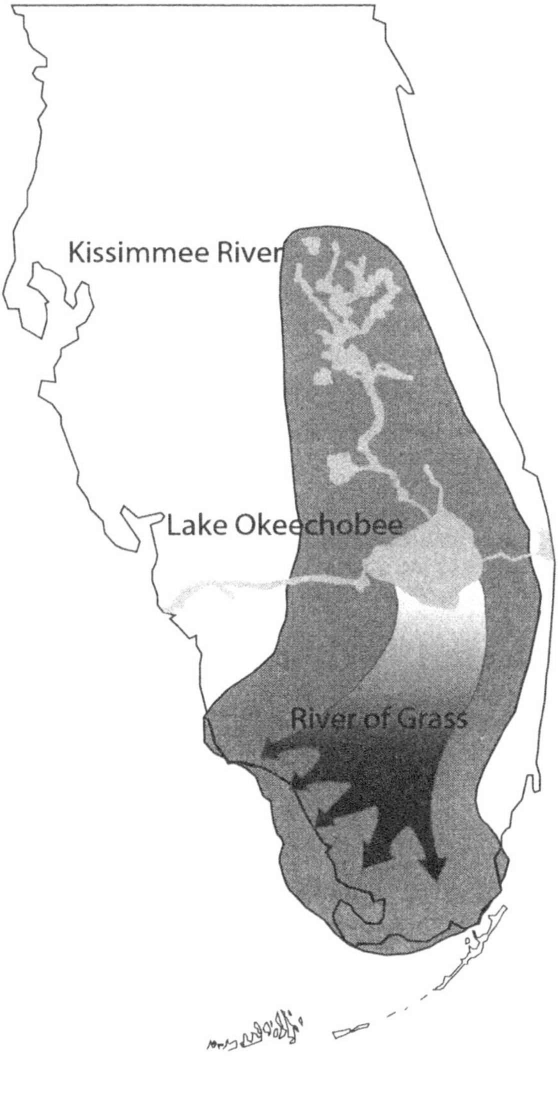

Figure 6.1 Pre-drainage Water Flow in South Florida

southeast coast. Although the Indians occasionally strayed inland, the wetlands of the Everglades provided little in the way of permanent habitation. The Indians of South Florida were mostly hunters and gatherers that lived off the land eating shellfish, deer, small animals, turtles, fish, and wild plants. They took advantage of the abundant fresh and salt water resources and prospered because they were adapted to the environment of the Everglades. The arrival of the Spaniards proved disastrous for the native peoples. Warfare and disease almost completely wiped out the population and when the Spaniards left Florida in 1763, the handful of native American survivors left with them.

The Engineered Everglades

When the Europeans arrived in Florida in the nineteenth century, they encountered a hostile landscape with few people. The original Indian tribes had disappeared with the Spaniards and the area was now inhabited by Seminole and Miccosukee Indians who had been forced to move to Florida from Georgia and Alabama. In contrast to the original tribes, the Miccosukee and Seminoles lived on tree islands in the interior of the Everglades as well as in the Big Cypress and Thousand Islands areas.

The climate and the organic soils of the Everglades encountered by the first Euro-Americans led them to believe that the area could support a lush tropical agriculture and by the late 1800s the first efforts were underway to drain the worthless swamp and turn it into productive agricultural lands. The first drainage canals were dug in the upper Kissimmee River and between Lake Okeechobee and the Caloosahatchee River. In 1903, the construction of the Miami River marked the beginning of a series of canals that connected the northern Everglades directly with the Atlantic Ocean. Construction of these canals made it possible to divert water from the original River of Grass and enabled the development of agriculture south of Lake Okeechobee. By 1921, about 2000 people had settled in 16 settlements in the Lake Okeechobee area (U.S. Army Corps of Engineers and South Florida Water Management District, 1999).

During the first reclamation efforts, the new residents of South Florida were repeatedly confronted with the natural forces in the area. In 1928, a total of 2,400 people drowned in South Florida, after a hurricane caused Lake Okeechobee to overflow (South Florida Ecosystem Restoration Task Force, 2002). In March of 1943, a large part of the Everglades went up in flames after a severe drought and a period of frost (Zaneski, 2001). In October 1944, a ten months long drought began which again caused severe fires (Zaneski, 2001). In 1947 two hurricanes caused a disastrous flood, flooding 90 per cent of Southeast Florida for six months (Light and Dineen, 1994). Faced with continuous threats to South Florida residents, the federal government intervened. In 1948, Congress approved the Army Corps of Engineers Central and Southern Florida Project. The goal of the project was to build a water management system that provided flood control and water for agricultural, municipal and industrial uses, prevented salt water intrusion and protected Everglades National Park and other fish and wildlife sources. The Corps started with construction of the Central and Southern Florida (C&SF) Project in

1950 and completed the Project in 1970. The finished Project consisted of about 1,000 miles of canals, 712 miles of levees, 150 control structures and 16 major pump stations (U.S. Army Corps of Engineers and South Florida Water Mangement District, 1999).

The C&SF Project had a significant impact on large portions of the original Everglades. The Kissimmee River was channelized and Lake Okeechobee was diked to prevent uncontrolled overflows from the lake. Immediately south of Lake Okeechobee, the area now known as the Everglades Agricultural Area, was drained and ground water levels were managed to reduce flood damages to agricultural production. Further south on the lower east coast, a drainage system was constructed to allow for urban, suburban and agricultural development. Finally, central portions of the Everglades were diked to create the Water Conservation Areas, which were intended for urban flood control, water conservation, the prevention of salt-water intrusion, recreation, preservation of fish and wildlife, and water supply for Everglades National Park. Although it was expected that the Water Conservation Areas would play some role in the preservation of fish and wildlife, the only area intended for preservation in its natural state was Everglades National Park, which was authorized by Congress in 1934 and established in 1947. The park encompasses 2353 square miles and was established to protect the unique tropical biological resources of the southern Everglades ecosystem (U.S. Army Corps of Engineers and South Florida Water Mangement District, 1999).

The C&SF Project effectively drained half of the original Everglades which allowed extensive urban development of the lower east coast as well as the establishment of agriculture in the newly created Everglades Agricultural Area. Urban development in south Florida originally started along the Atlantic Coastal Ridge where the land was slightly higher in elevation. The C&SF Project made it possible for urban development to expand westwards toward the Everglades and South Florida's population grew dramatically from 500,000 people in the 1950s to today's population of about six million people. The C&SF Project allowed for the development of the area with a total economic output of over $200 billion annually with a booming $14 billion tourist industry (World Resources Institute, 2000), a $2 billion agricultural sector (World Resources Institute, 2000) and a $70 billion a year international trade sector (Enterprise Florida, 2000).

However, the C&SF Project failed to protect Everglades National Park and for that matter it failed to protect all environmental resources in South Florida. In order to provide flood protection the C&SF Project effectively dumps 1.7 billion gallons of water per day into the Atlantic Ocean in the east and the Gulf of Mexico in the West, leading to water shortages in the Everglades (U.S. Army Corps of Engineers, 1998). In recent years, increased urban development and the resulting water runoff has made it difficult for the managers of the existing water system to handle the natural fluctuations in rainfall. The extremely wet winter of 1997, for example, flooded large areas of the Southern Everglades, caused high water levels in Lake Okeechobee, killed fish in the St. Lucie River and flooded the nesting area of the endangered Cape Sable Seaside Sparrow in the southern Everglades. The following dry summer of 1998, on the other hand, led to massive wild fires. In the spring of 2001, continued dry conditions forced the South Florida Water Management District to issue Phase Two water restrictions, which seriously limited the outdoor water use for south Florida's residents. Figure 6.2 shows the current flow of water in South Florida. The figure shows that most water that historically flowed into the River of Grass from Lake Okeechobee, is now diverted into the Atlantic Ocean and Gulf of Mexico through the St. Lucie and Caloosahatchee rivers.

The release of large quantities of fresh water has caused serious ecological problems in the salt water estuaries of the St. Lucie and Caloosahatchee rivers, such as fish lesions and a decline in estuarine productivity. It has also reduced the amount of water that flows into Everglades National Park and Florida Bay by 70 per cent which has lead to excessively high concentration of salt in the Bay. In recent years, Florida Bay has seen fish kills and a reduced pink shrimp and Florida lobster catches. Mangroves, seagrass, and sponges have died-off and there has been a general decline in estuarine productivity. Finally, this quick dispersion of water into the ocean is depleting the fresh water supply in South Florida by reducing the percolation of water into the aqulfer.

In addition to the problems with the water quantity, water quality has deteriorated over the past 50 years. Urban and agricultural run-off have deposited large quantities of phosphorous and nitrogen, causing eutrophication of the Everglades, and Florida and Biscayne Bays. The problems with water quality are exacerbated because of the destruction of wetlands that acted as natural filters and retention areas. As a result, many water bodies throughout south Florida presently do not meet water quality standards for phosphorus and nitrogen. The Everglades also suffer from high mercury concentrations which has lead to fish consumption advisories for most of the Everglades waters and the further demise of the Florida Panther.

It is clear that the C&SF project has changed most of the characteristics of the predrainage ecosystem. Besides the problems with water quantity and water quality, the mere conversion of half of the original Everglades ecosystem to agricultural and urban uses has tremendous consequences. The loss and alteration of habitat has resulted in a dramatic decline of native species and the prolific spread of exotic species. According to recent estimates there are just 50-60 Florida Panthers left, wading bird populations have declined with 90 per cent and 68 plant and animal species in the Everglades are listed as endangered or threatened (U.S. Army Corps of Engineers and South Florida Water Management District, 2000). Meanwhile, Everglades National Park alone has been invaded by over 220 exotic plant species. The main problem species are Melaleuca, Brazilian Pepper, Australian Pine and Hydrilla, all of which are widespread and have replaced native species (South Florida Water Management District, 2001). Less visible is the reduction in the system-wide levels of primary and secondary production and changes in the proportions of community types within the remaining system.

Towards Restoration of the Everglades

The first problems with the C&SF Project occurred from 1963 to 1965, when the South Florida Water Management District (SFWMD) filled the newly constructed Water Conservation Areas with water and effectively prevented water from flowing

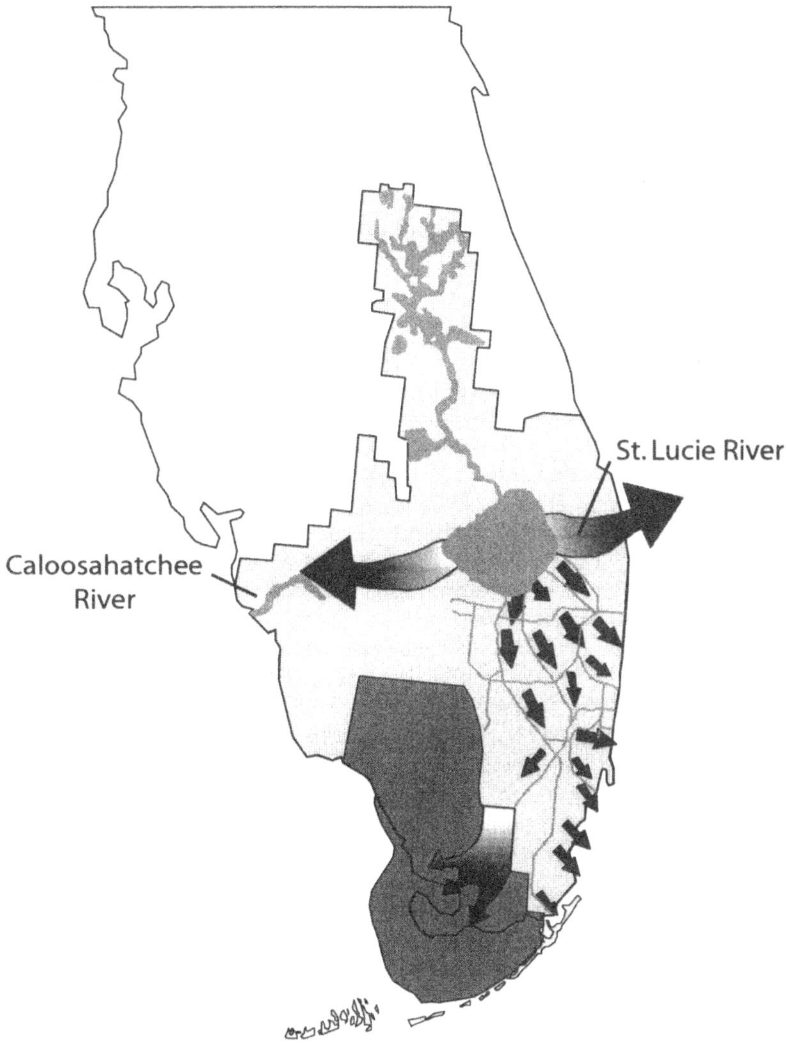

Figure 6.2 Current Water Flow

into Everglades National Park (World Resources Institute, 2000). The drought conditions that resulted from this caused extreme stress on wildlife in the park. Consequently, Congress ordered the SFWMD to deliver adequate amount of water to the park. Then, from late 1970 until the winter of 1971, South Florida was confronted with a severe drought. The drought caused muck fires in the Everglades and for the first time saltwater intrusion threatened the Biscayne Aquifer. Faced with the consequences of the drought, Governor Reuben Askew convened the Governor's Conference on Water Management in South Florida in August 1971. As a result of the Conference, the 1972 legislature adopted legislation that required local governments to adopt comprehensive plans by 1975 and ultimately lead to Florida's growth management program. In 1980 and 1981, a severe drought occurred again which lead to the initiation of the 'Save Our Everglades' Program by Governor Graham in 1983.

In 1992, after all the reviews of particular components of the C&SF Project, Congress directed the Army Corps of Engineers to review the entire Central and Southern Florida Project to determine if modifications in the system were necessary. In the 1996 Water Resources Development Act (WRDA), Congress mandated the Army Corps of Engineers to formulate a Comprehensive Review Study of the original Central and Southern Florida Project (The Restudy) and submit it to Congress by July 1, 1999 (Water Resources Development Act, 1996). The purpose of this Restudy was to develop modifications to the Central and Southern Florida Project to restore the Everglades and Florida Bay ecosystems while providing for the other water-related needs of the region (U.S. Army Corps of Engineers, 1998). The Central and Southern Florida Project Comprehensive Review Study Final Integrated Feasibility Report and Programmatic Environmental Impact Statement was completed in April 1999 and submitted to Congress on July 1, 1999. The first two pilot projects were authorized by Congress in the 1999 Water Resources Development Act. The complete Central and Southern Florida Project Comprehensive Review Study, now known as the Comprehensive Everglades Restoration Plan (CERP) was part of the 2000 Water Resources Development Act (WRDA, 2000), which was signed into law by President Clinton on December 11, 2001 (Water Resources Development Act, 2000). Title Vl, section 601 of the WRDA 2000, authorized the start of four pilot projects and ten initial projects for a total estimated cost of almost $1.4 billion. This leaves approximately 26 components, and $6.2 billion that will have to be authorized in WRDAs from 2002 to 2014.

The Florida Legislature passed the Restudy Bill in April 1999, among other things, this bill authorized the South Florida Water Management District to be the local sponsor for CERP. In May 2000, the Florida Legislature passed the Everglades Restoration Investment Act which committed the State of Florida to contribute over $2 billion to CERP, which fulfills the first ten years of the state's share of CERP. In that same month, the Florida Legislature passed the Lake Okeechobee Protection Program which appropriated $38.5 million for the restoration of Lake Okeechobee.

The Comprehensive Everglades Restoration Plan

The Comprehensive Plan encompasses approximately 18,000 square miles, stretches from Orlando to the most southern tip of Florida, includes 16 counties and is home to 6.3 million people. The plan is estimated to cost $7.8 billion dollars which will be equally split between Florida and the Federal government. The final stages of the plan will not be completed until 2050.

The goal of CERP is the recovery of healthy and sustainable ecosystems throughout South Florida while providing flood protection and fresh water supply for South Florida's residents. According to the Army Corps of Engineers and South Florida Water Management District (1999) the plan will:

- Improve the health of over 2.4 million acres of the South Flonda ecosystem, including Everglades National Park
- Improve the health of Lake Okeechobee
- Virtually eliminate damaging freshwater releases to the estuanes
- Improve water deliveries to Florida and Biscayne bays
- Improve water quality
- Enhance water supply and maintain flood protection

The idea behind the plan is to recreate part of the original water flow through South Florida as was shown earlier in figure 6.1 and create a buffer between the urban area and the remaining Everglades. Figure 6.3 shows the flow of water once the comprehensive plan is finished. Although the total Everglades ecosystem is dramatically reduced compared to the pre-drainage situation (less than 50 per cent), the water flow is similar. Water is no longer discharged into the ocean through the St. Lucie and Caloosahatchee rivers but slowly finds its way from Central Florida to Florida Bay and the Gulf of Mexico. To achieve this, 14 large Surface Water Storage Resevoirs (SWSRs) will be built between the urban area on the southeast coast and the Everglades ecosystem. These reservoirs will have the capacity to store 1.5 million acre-feet of water and encompass an area of approximately 181,300 acres (U.S. Army Corps of Engineers and South Florida Water Management District, 1999). The intent of these reservoirs is to store water during wet periods that can then be released during dry periods. The second feature is the creation of seven Storm Treatment Areas (STAs) that will cover 35,600 acres. These STAs will treat water from urban and agricultural runoff before it is released in the natural areas and are in addition to the 44,000 acres that are already under construction under the Everglades Forever Act. The main function of the STAs is to remove phosphorus from the water before it is released into the Everglades watershed. Additionally, two wastewater treatment plants will be built in Miami-Dade County, which will be capable of treating 220 million gallons of waste water a day. These two plants will make it possible to discharge treated waste water from Miami-Dade County into the Everglades ecosystem and Biscayne Bay. The most controversial component of CERP is the establishment of more than 300 wells that will allow pumping of 1.6 billion gallons of water a day into the upper Floridan Aqulfer.[1] This water will then be pumped to the surface when

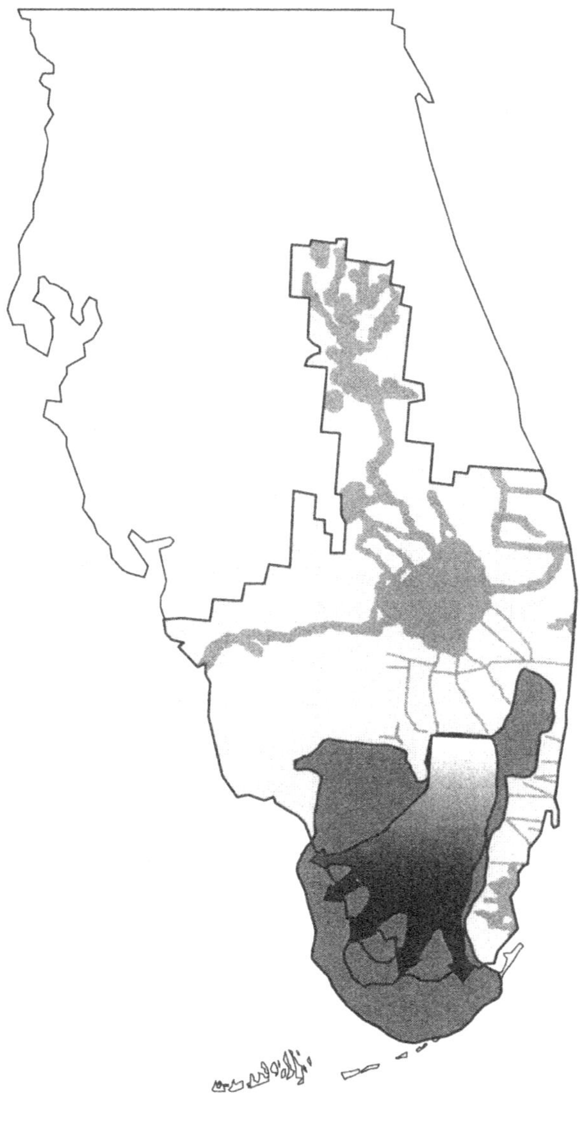

Figure 6.3 Water Flow After CERP is Complete

needed. In order to reestablish the historical north-south sheet-flow of water through the Everglades, 240 miles of canals and levees will be removed.

Potential Problems with CERP

CERP has been heralded as the largest water project in the history of the U.S. With its $7.8 billion price tag and its extensive modifications to the existing system this is probably true. The plan will undoubtedly improve water quality and restore at least some of the original water delivery to the natural system. However, there are many uncertainties about how well the different components of CERP will function. The main uncertainty that has been brought forward over and over again involves the deep water storage areas. Deep water storage has never been done on the scale proposed in CERP and both the feasibility and the effects on the aquifer are unclear. At the same time, deep water storage is a critical component of CERP and it is questionable that the system will function properly without it.

Both the Corps of Engineers and the SFWMD have been optimistic about the deep water storage systems. As a response to uncertainties they have adopted a so called 'adaptive assessment approach'. According to CERP:

> Adaptive assessment is a process which involves the iterative use of models, research, and monitoring in conjunction with on-going planning, in order to revise, improve, and fine tune management procedures. It involves an iterative process of developing management tactics, and provides a process for breaking impasses where agencies are unwilling to proceed because of an inadequate knowledge base (U.S. Army Corps of Engineers and South Florida Water Management District, 1999, pp. 5-32).

In plain English this means that several pilot projects will be constructed that will provide more knowledge about the functioning of some of the components of the plan. Based on the experience with a pilot project, a particular component will then be adopted or changed. Adaptive management gives the plan flexibility and allows Everglades restoration to move forward even though all the information and knowledge is not available yet. At the same time, it adds uncertainty about the functioning of the entire system and it adds uncertainty about the real ultimate price tag.

The second issue involves the Everglades agricultural area. As can be seen from Figure 6.3, CERP effectively cuts the original Kissimmee-Okeechobee-Everglades watershed in two parts with the Everglades agricultural area, just south of Lake Okeechobee, as the barrier between the northern and southern part of the watershed. According to representatives from the Army Corps of Engineers, the reason for this division is that soil subsidence in the Everglades agricultural area has created a system in which water no longer naturally flows from Lake Okeechobee southwards, but instead flows northwards into the agricultural areas. Undoubtedly, the political power of 'Big Sugar' that operates most of the agricultural lands in the agricultural area also played a major role in this component of the plan.

The third issue is public involvement. In a project of this size, public involvement and support is critical. It seems that both the South Florida Water Management District and the Army Corps of Engineers are aware of this and there have been a relatively large number of public meetings about CERP. After the formulation of the Draft CERP, the SFWMD and the Corps held a series of 12 public meetings about the plan. The meetings were held in different locations and attended by both the executive director of the SFWMD and the Chief Engineer of the Corps. Since then, there have been meetings for specific projects of CERP such as the Economic and Environmental Justice Plan and the Public Outreach Plan. Although there have been many meetings, most of which were well attended, there are some serious problems with the public involvement effort so far. The first series of twelve hearings seemed more like a public relations effort to garner the support of the public in South Florida. The hearings were set up the same with an Open House from 5 to 7 pm and a public meeting at 7 pm. The Open House consisted of a series of displays that explained the components of the plan with people from the SFWMD and the Corps at hand to answer questions. The public meeting part consisted of a brief introduction by representatives from the Water Management District and the Corps, after which members of the public had two minutes each to comment on the plan.[1] The comments were recorded and can be found in an appendix of CERP. Questions that people asked during the public meetings were not answered during the meetings but both the questions and the answers can also be found in an appendix of CERP.

According to CERP, public outreach consists of two components, involvement and information. The experience with public outreach so far has been that the meetings for the general public have focused on the information component and have not allowed for meaningful involvement from the audience.[2] CERP states that the involvement of the public in further planning and design will remain a priority during the implementation of the program. Given the content of public outreach so far, it is questionable that this involvement will entail more than one way information provision. CERP distinguishes between program and project level public involvement. With program level activities CERP refers to the entire plan, while project level activities refer to specific components within the plan. While each project will have its own public involvement and outreach requirements and activities, CERP acknowledges that there will be a continuing need for program-level outreach efforts. The primary objectives of the program-level public outreach activities are twofold: to keep the public informed of the status of the overall program and the key issues associated with its implementation; and provide effective mechanisms for public participation in further plan development.

CERP and Population Growth in South Florida

Besides the direct consequences of CERP, there are secondary effects that have either not been addressed or have been addressed poorly. Among the most important of these secondary effects are the consequences for the location of population growth in South Florida and consequences for existing urban areas. The

16 counties that are included in the Comprehensive Plan have a current population of about 6.3 million people and are projected to grow to 11.5 million by the year 2050 (U.S. Army Corps of Engineers and South Florida Water Management Management District, 1999). Most of the population (approximately 63 per cent) in the study area lives in the lower east coast. Another 20 per cent of the population lives in the Kissimee area, just south of Orlando. The future population projections until 2020 that have been used for CERP are based on extrapolation of the 'medium' estimate projection of the Bureau of Economic and Business Research at the University of Florida. Using the same calculations for the 'low' and 'high' projections the population in the study area can be projected to be anywhere between 7.2 and 14.1 million residents. Although it has been common practice for planning agencies in Florida to use the medium population projections, it seems that CERP should have also considered the consequences of the low and high projections for South Florida. The current problems with water in South Florida are at least partly caused by the fact that the original Central and Southern Florida Project was designed to accommodate the water needs of a projected population of 2 million residents in South Florida, not the current 6 million. It seems prudent that for a massive project, such as The Restudy, different population scenarios would be used.

Historically, the BEBR medium population projections have been a fairly good predictor of real population growth in South Florida but it is unclear if it will do as good in the future. Until recently, land for new development was readily available and population growth was almost simply a matter of how fast developers could build new subdivisions. Now, the tide is changing, several counties in the lower east coast are close to complete build-out. To accommodate further population growth, these counties will not only have to redevelop sites within the existing urban pattern, they will also have to increase population densities substantially. For example, in the built-up area of Broward County the population density in 1995 was 5.2 people per acre. Under the medium population projection this will have to increase to 7.5 people per acre in 2020 and according to CERP to 7.8 people per acre in 2050. It is clear that this means higher density development. Current residential developments in the suburbs however are typically on large lots with large homes with densities that are around 3 people per acre. It is questionable whether Broward County's population will continue to increase at the medium projected growth rate once the only available new housing will be in the form of relatively high density near the urban core. The same argument can be made for other counties on Florida's lower east coast. Of course this does not necessarily mean that population growth in South Florida as a whole will be substantially less. The Upper Kissimmee Basin, including Orlando and Disney World, is developing rapidly, especially in the outer urban fringes on former citrus lands (Governors Commission for a sustainable South Florida, 1995).

Current growth patterns also indicate significant growth on Florida's Southwest coast. With enough land available on the fringe of the Everglades and no established urban growth boundary it seems a likely place for new, low density suburban sprawl into the Everglades. The problem with CERP is that the plan only looks at the Southeast coast of South Florida. The Plan states that:

> There are additional water resources problems and opportunifies in southwest Florida requiring studies beyond the scope of The Restudy recommended Comprehensive Plan. In this regard, a feasibility study for Southwest Florida is being recommended to investigate the region's hydrologic and ecological restoration needs. (U.S. Army Corps of Engineers and South Florida Water Management District, 1999, p. X).

The lack of consideration of Florida's southwest coast is a serious short coming, especially since the comprehensive plan tries to restore the traditional water flow from Central Florida to the south and Southwest coast. Under the current plan it would be possible and, given the current fast development of cities like Fort Meyers and Naples, likely that west coast suburbs will encroach on the Everglades ecosystem from the north and northwest. Since the plan does not envision a buffer zone, in the form of surface water reservoirs or stormwater treatment areas, between the west coast and the Everglades ecosystem, this could potentially cause problems with both increased water run-off and increased pollution in the southwestern Everglades, Big Cypress National Preserve, Florida Bay and the Gulf of Mexico. It is likely that new development in Southwestern Florida will also lead to increased salt water intrusion in that area.

CERP and the Existing Urban Area

CERP pays surprisingly little attention to existing urban development. The seemingly obvious link between restoration of the Everglades and urban development patterns in South Florida is only casually referred to as one of the reasons for the current problems with the Everglades Ecosystem. The lack of an 'urban' or 'land use' component in CERP was brought up several times during the initial 12 public hearings. However, the final plan still fails to address the issue. In one of the Restudy Updates published by the Corps and the SFWMD, this shortcoming was put aside by stating:

> ... it is the statutory responsibility of the South Florida Water Management District to plan for future water supply and management needs. Limiting growth is generally a local government issue (Governor's Commission for a Sustainable South Florida, 1999, p. 14).

Based on the agencies that have been involved, the lack of an urban or 'land use' component in the Restudy is not surprising. Table 6.1 shows the affiliation of the people that were involved in the preparation of the comprehensive plan. It is clear that the project was dominated by the South Florida Water Management District and the Army Corps of Engineers. At the same time there was no representation from any of the five Regional Planning Councils. Nor was there involvement from local government or any land-use agency.

Besides the previous discussed problem with a possible shift in population growth form Florida's lower east coast to the southwest coast, CERP ignores the consequences of the restoration for the existing urban area. Since the 1950s, urban growth in southeast Florida has been in the form of steady expansion from the coastal area westward towards the Everglades. This growth has largely been in the form of low density urban sprawl, made possible by the construction of major east west highways. As mentioned before, development in the lower east coast will not be able to continue its westward stride much longer. Instead developers will increasingly have to look at infill and redevelopment within the existing urban area. Most of the redevelopable land in the lower east coast is located in the urban core, where there is a relatively high percentage of low-income and minority residents. Recent success stories, such as Victoria Park and downtown in Fort Lauderdale and downtown West Palm Beach, show that this typicaily leads to an increase in property values and displacement of the original residents.

Table 6.1 Affiliation of members of plan preparation team

Agency	Participants
South Florida Water Management District	63
U.S. Army Corps of Engineers	
National Park Service	15
U.S. Environmental Protection Agency	5
Other Federal Agencies	18
State Environmental Agencies	8
County Environmental Agencies	
Academics	
Tribes	3

Infill and redevelopment will have to occur at higher densities than are now common in South Florida. These higher densities are likely to cause increased (polluted) urban run-off not only to the Everglades but also to the few remaining wetlands and natural areas within the urban area. At the same time infill will reduce the amount of open space and pervious surface in the urban area. CERP completely ignores the potential consequences of increased densities and a reduction in pervious surface for Everglades Restoration, nor does it acknowledge a possible role of wetlands and natural areas within the existing urban area. This shortcoming was also noticed by the Governor's Commission for a Sustainable South Florida (1999):

> The successful restoration of Everglades functions is dependent not only upon the establishment of correct hydropatterns within the remaining Everglades but also upon the preservation and expansion of wetlands: including those within urban areas that once formed the eastern Everglades...The SFWMD and the Corps should acknowledge the important role of urban natural areas as an integral part in the restoration of a functional Everglades system (Governor's Commission for a Sustainable South Florida, 1999, p. 15).

Conclusion and Discussion

Given the tremendous changes that have occurred in South Florida over the past 119 years, it is impossible to recreate the pre-drainage Everglades. The restoration goal for CERP therefore is to create a 'new' Everglades, one which will be different from any system that existed in the past, and one which will be substantially healthier than the current system (U.S. Army Corps of Engineers and South Florida Water Management District, 1999, pp. 5-36).

CERP argues that the restoration will be successful if it recovers the ecological components and patterns which characterized the pre-drainage system. One can wonder if it is realistic to assume that a viable system will exist after CERP has been implemented. The original Everglades could only exist by virtue of its large scale. The large scale gave the system its resilience and provided habitat for top predators and niche specialists. Even when, by 2050, CERP is completely implemented and all its components work both individually and as a collective, the 'new' Everglades will at best occupy half of its original size. At the same time, according to current projections, the human population will have doubled or even tripled.

It is interesting that Florida's growth management program started with the Governor's Water Conference in 1971. Although there was general concern about the negative consequences of growth, it was water supply that acted as the initiator. Now, 30 years later, it is again water that acts as an important factor in growth management in South Florida. This time it is both water quality and water quantity. There is another big difference between the developments of 1971 and CERP; where the 1971 Water Conference clearly acknowledged that water and land in South Florida are inextricably linked, CERP does not acknowledge this link. Instead CERP focuses completely on water and ignores the consequences of water management for urban development and population growth. CERP does not discuss the relationship between the restoration of the Everglades and regional and local land use plans, nor does it discuss the possible consequences for population growth in South Florida. Unfortunately, the plan does not acknowledge its role in shaping future urban development and the direction of future growth in South Florida. Instead, the plan insists on being a water management plan that has no connection with urban development, growth management or population growth.

Even with all the shortcomings mentioned in this chapter, CERP offers hope for an ecosystem that was on its way to total destruction. It is debatable whether CERP will recreate an ecosystem that resembles the original ecosystem. Given the reduction in size, it is even not certain if the plan will lead to a system that can sustain itself. However, for the first time, the people of South Florida have acknowledged that the Everglades are not a wasteland and that it forms the basis of the success of the region. That by itself might be worth the hefty price tag!

Notes

1 Personal observation of author during public meeting for CERP on November 10, 1998 in Fort Lauderdale and November 12, 1998 in West Palm Beach.

2 Statement based on personal observation during public meetings for CERP on November 10, 1998 in Fort Lauderdale and November 12, 1998 in West Palm Beach, the public hearing on the Economic and Environmental Justice Plan and the Public.

References

Enterprise Florida (2000), website. www.eflorida.com.

Governor's Commission for a Sustainable South Florida (1995), *Report of the Governor's Commission for a Sustainable South Florida,* Tallahassee.

Governor's Commission for a Sustainable South Florida (1999), *Report on The Draft Implementation Plan of the C & SF Project Restudy*, Tallahassee.

Light, S.S. and Dineen J.W. (1994), 'Water Control in the Everglades: A Historical Perspective,' in S.M. Davis and J.C. Ogden (eds), *Everglades: The Ecosystem and Its Restoration,* St. Lucie Press, Delray Beach, pp. 48-68.

McCally, David (1999) *The Everglades: An Environmental History*, The University of Florida Press; Gainsville.

South Florida Ecosystem Restoration Task Force (2002), *Biennial Report for 2001-2002 of the South Florida Ecosystem Restoration Task Force*, River Ranch, Florida.

South Florida Water Management District (2001), *2001 Everglade Consolidated Report*, West Palm Beach.

U.S. Army Corps of Engineer (1998), *Overview: Central and South Florida Project Comprehensive Study Review*, Jacksonville.

U.S. Army Corps of Engineers and South Florida Water Management District (1999). *Central and Southern Florida Project Comprehenisve Review*, West Palm Beach.

U.S. Army Corps of Engineers and South Florida Water Management District (2000), *Rescuing an Endangered Ecosystem; The Plan to Restore*, West Palm Beach.

Water Resources Development Act (1996), P.L. 104-303.

Water Resources Development Act of 2000, P.L. 106-541.

World Resources Institute (2000), *World Resources 2000-2001: People an Ecosystems, The Fraying Web of Life*, World Resources Institute, Washington.

Zaneski, Cyril T. (2001), ' Anatomy of a Deal,'*Audubon*, July-August, pp., 49-53.

Chapter 7

Unorganized Maine: Regional Planning Without Local Government

Andrew Fisk

Epigram

'Attention of the State of Maine has frequently been called to the so called wild lands of the State. It is a matter of great importance and one upon which from time to time much has been written'. 1908 Report on the History of Wildlands of Maine.

Location and Nature of Jurisdiction

Maine's Land Use Regulation Commission (LURC) governs a vast and varied geography with over 400 unincorporated townships, 39 organized towns and 300 coastal islands, lands commonly referred to as the 'unorganized territories'. (Figure 7.1) The unorganized territories include over 3,000 lakes, hundreds of miles of rivers including the St. John, the longest free-flowing river on the East Coast, as well as the Penobscot River, which drives the largest privately owned hydroelectric system in the United States. With over 10.4 million acres and a year-round population of just 12,000 the unorganized territories support a population density of one person per square mile, an allocation more reminiscent of the Great Plains than the crowded Eastern Seaboard. LURC's jurisdiction is in many ways simultaneously a very big place and a very small place.

In between the infrequent resident lies commercial forest. This timberland, which covers close to 17 of Maine's 19.7 million acres of land, has remained in continuous (and contentious) management for the past two centuries. While the forest products industry is far and away the largest industry within LURC's jurisdiction, tourism has become more important over the last thirty years as white-water rafting, snowmobiling, and other outdoor adventure recreational businesses have begun to develop alongside the more traditional hunting, fishing, and vacationing at seasonal lakeside homes, or 'camps'.

After a period of predictable controversy in the 1970s and 1980s over the tyranny of state-mandated land use regulation, and then an equally predictable détente between regulator and regulated, a great deal is now changing in the 'Great North Woods'. New and economically robust recreational industries are moving inland from the coast and are growing up alongside an increasingly mechanized forest products industry,

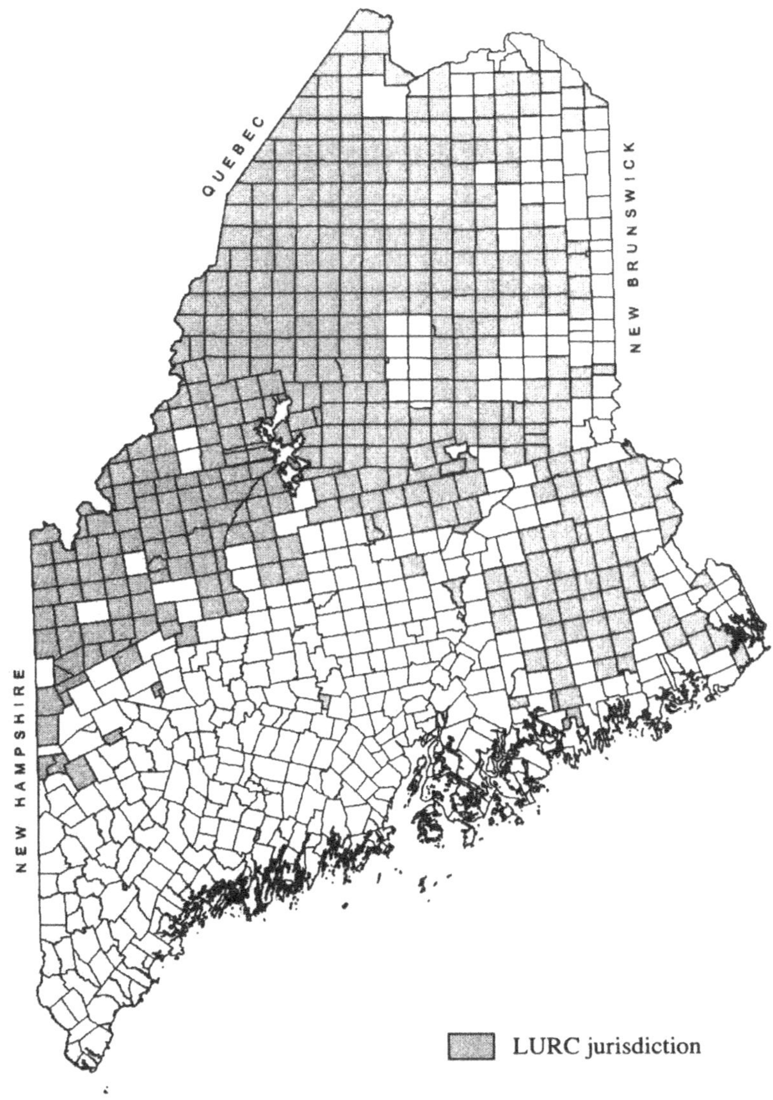

Figure 7.1 Maine Land Use Regulation Commission Jurisdictional Coverage

which in turn faces increased global competition and internationalization. The long-term owners of mills and timberlands have sold to institutional investors from other countries and other regions of the U.S. who are a breed apart from the paternalistic and locally-controlled firms of prior years. The nature of tenure on the timberlands is changing on a wide scale as landowners, both new and old, are selling development rights on millions of acres to ensure these lands are managed for timber in perpetuity. And while frequentlypricey recreational camp lots continue, commonly called 'Kingdom lots', many counties continue to see population losses and only moderate economic growth. Maine, and in particular its more rural counties, remains at the bottom of the per capita income rankings of all the United States.[1]

History and Patterns of Land Ownership

'We bought Maine from Massachusetts in 1820, and we've been selling it back to them an acre at a time ever since'. – Anonymous wag

Maine's current landownership pattern holdings controlled by families, investment firms, and mill operators was cast as early as the 17th century when the State was a province of colonial Massachusetts. Vast private landholdings and a small population have been facts of Maine's existence since before the Revolution and have remained so into the 21st century. Indeed, 10 firms own approximately 75 per cent of the land base in the unorganized territories.[2]

But these ownership patterns did not arise quickly or certainly. In Maine's early years of European settlement there were no land rushes or Oklahoma-like 'Sooners' racing across the border. Aside from a few very large acquisitions, such as William Bingham's purchase in 1793 of two million acres around the Kennebec River, land went begging for many decades in the 18th and 19th centuries. Following the revolution, the State of Massachusettssought to divest as much of its holdings in the Province of Maine as it could in order to pay war-related debts. The State proposed a lottery sale in 1786 of 50 townships after its three-year-old land office had sold only a marginal number of lots directly. The State hoped to sell 2,720 tickets at 60 pounds a ticket, but only some 400 were sold, leaving the winners with 165,280 acres and the State netting on balance 52 cents for each acre. Because cash was short and land up north was ample, provincial Maine acreage was used to pay Revolutionary war widows in lieu of pensions as well as even state-supported ministers.

When the Federal Congress issued Maine a grant of independence as a State, a significant amount of the public land remained in common ownership between the two states, leading to no small amount of friction over its improvement and method of disposition. But by 1878 all eight million acres of public land had been alienated with tracts being given to colleges and institutions, auctioned to individuals, or sold to investors including a 700,000-acre grant to a railroad syndicate. Some tracts were set aside as school 'public' lots that would eventually be sold to raise revenue when townships became settled.

The availability of land allowed for many different family ownerships to reach notable size throughout the 19th and 20th centuries. To date however, only a few have remained intact and in active timber management: the Pingree family holdings are one. After making a comfortable living as a merchant ship owner in Salem, Massachusetts in the 1830s and 40s, David Pingree sent agents north to the State of Maine to acquire as much timber land as could be bought. Captain Pingree like the Dunns, the Prentisses, the Carlisles, and the Webbers would do as well and eventually assembled holdings of close to one million acres throughout the northern and western counties. The Pingree ownership has remainedin common and undivided ownership among what has now grown to over 160 family members. The family's 960,000 acres is among the country's largest and oldest private family holdings.

Family holdings owned for the long-term are now the exception, not just in the nature of their ownership, but their duration. Commercial timberland is now trading hands at rates not previously seen. Within the space of approximately three years (1997-2000) almost 20 per cent of the State's 19 million acre land base traded hands as timber companies and mill owners divested holdings and new timber companies or investment groups moved into the State (Young, 2000). In addition the increasingly regular purchases by wealthy business executives of 15,000 or 20,000 acre recreational kingdom lots at unusually high prices for land generally sold at timber values of $200 - $300 an acre and has prompted as much public interest as the larger commercial timberland sales.[3]

The sale of kingdom lots has prompted a great deal of speculation and legislation about public access for recreation. Historically, extensive ownerships have fostered a longstanding tradition of open lands and free access. Many large landowners have not and still do not restrict access to their lands, reachable on an extensive private road network. The landowners allow camping, fishing, and hunting to all comers. There is widespread public concern that these kingdom lot owners, who are generally keenly interested in their privacy, will post no trespassing signs on their substantial holdings and begin to chip away at Maine's proudly and vigorously held tradition of free public access. Where much of the jurisdiction has free public access, there have been some successful efforts to institute a user fee system in order to recoup the not insignificant costs of free public access, including managing campsites and trash disposal. Over the last thirty years several large landowners have created a coordinated program of managing recreational access to several hundredthousand acres of lands through a user fee system administered at gate checkpoints. While the gate system is widely accepted, significant and successful public protest has arisen when gate or user fees rise out of step with what users find appropriate, as one new landowner (an out-of-state investment group) recently discovered to their surprise. A sharp increase in fees provokedpublic protests, widespread media attention, numerous pieces of proposed legislation, and a legislative taskforce.

Lease Lots

Land tenure in Maine also includes a surprising number of lots leased to individuals and commercial guides for hunting and fishing cabins. There are several thousand recreational lease-lotsheld by a number of the large landowners and the State. It is not uncommon to find small lease lots throughout the North Woods that date to the early part of the century and have been held across several generations of families. Most of these lease lots were created by paper and timber companies earlier in this century for a variety of reasons, not the least of which was good will. The large landowners, many of which prior to the recent round of land sales were also the mill owners and direct employers of thousands, offered leases to employees at very economical rates as a form of bonus. Until recently these leases were very inexpensive, with an average annual payment for a small waterfront lot being only a few hundred dollars.

Whether these lots were issued by the industrial landowners and employers or the private family holdings they all provided access to the woods and lakes for the average Mainer. And while the tenure was never secure in a legal sense – leases were renewed yearly, but in effect carried a 90-day termination notice – leaseholders felt secure that the land base itself was stable and so the owners had little reason to cancel leases. Particularly since landowners had sound reason to cultivate good relations with leasees as an important source of political goodwill and employee loyalty.

The mutuality of the leasee/leasor relationship has nurtured probably apocryphal stories where a company could be guaranteed five hundred ready letters or phone calls in a political crunch for the price of keeping land management, popularly known as tote roads, plowed early and often. Some folks will claim the clout of these leasees and their letters was such that tote roads were often plowed in advance of the active logging roads. The practice of leasing, rather than selling land in fee, was advantageous to the landowners. It provided them with flexibility where leases could be terminated, although in practice cancellations are infrequent, and allowed them a measure of control over what could become incompatible development and demands for services. For instance, to this day many hundreds of leases prohibit pressurized water, electric service drops, and permanent occupancy. Because these leased lots are situated along active logging roads the landowner does not want them to become overly suburban and the leasees so develop an animosity to timber harvesting operations occurring in their backyards. Strictures and insecurity of tenure aside, lease lots are readily accepted by Mainers as a sound and inexpensive investment in a piece of the Great North Woods. It is unclear how long leasing will continue as a land management strategy, as several of the new commercial timberland owners have begun to market several hundred lease lots in the last two years, with more expected to be put on the market.

The State is also an active leasor of recreational camp lots. As a result of land swaps, consolidations, and sales in the early 1970s, the State leases some 200 lots throughout the jurisdiction in a low-key, but politically popular program. With some exceptions all the leases now managed by the State were created when the land was privately held; they were then grandfathered by statute when the State took ownership. The statutory protection of the lease program was enacted to reassure leaseholders that they would not be evicted. Although the State has gradually accommodated itself to being a recreational landlord, it does not find the situation ideal. The State Bureau of Parks & Lands has not created any additional leases since the program began, and where possible has sold lots to leasees where they were located along public roads or at the fringes of holdings. However, many of the most coveted and strenuously protected lease lots are remote in-holdings in pristine locations which if sold, the State feels, would adversely impact the broader public uses of the adjacent public land. On balance however, both leasors and leasees are satisfied with the status quo, particularly compared to private leases, where State leases have five-year terms, lower fees, and higher thresholds for cancellations. Any lease terminations by the State require approval of the full Legislature, which is highly unlikely (Austin, 2000).

Formation of the Commission

The idea for a regional planning entity in the North Woods was first introduced as legislation in 1967, and 'An Act to Create the Wildlands Use Regulation Commission' was successfully voted out of committee (Senator Paper 506, 1967). However, opposition by the large landowners caused the bill to be indefinitely postponed in the full Senate. After this first effort was derailed, the Legislature directed its research arm to study the desirability of controlling the use of land and water resources in the unorganized territories. What became known as the 'McKee Report' after its chairman John McKee, was issued in January 1969 after six months of meetings by a select committee that reviewed the original legislation.[4] The McKee Report did not stray far from the original legislation passed by the committee in 1967, and in fact it became more stringent as language was clarified.

Aside from the broader initial controversy over whether planning and land use regulation were even needed in the North Woods, the most debate and discussion was focused on which lands were to be regulated and to what extent. The Commission's original organic act, which was passed in 1969, regulated only those areàs that were 500 feet from the traveled edge of any access road, lake, pond, river, or stream (House Paper 171, 1969). Remote water bodies were excluded from the proposal, where remoteness meant no part of a water body was within 500 feet of any access road. In short, the original legislation only included areas that were immediately facing development pressure and did not envision planning for areas which would in short order become accessible and desirable for development. Legislators were not concerned about the remote waterways because of the decline in the rivers' use for log drives and the creation of 'tote' roads for hauling timber to mills. The last log drives in Maine occurred in the middle 1970s, with concurrent increases in the number of miles of tote roads being constructed. These roads, which now run to more than 25,000 miles, have opened vast areas of the jurisdiction to four- and two-wheel drive vehicles in the last thirty years. These improvements in access have significantly improved recreational opportunities, however, they have also significantly sharpened the debate about the use of the woods – with many wanting more access that is more secure, while others argue these roads are the development inducing. So, under close public scrutiny, large landowners who now must be more sophisticated land managers of multiple uses are left sorting through the implications, costs, and benefits of these roads.

The initial political compromise with the forest products industry that constrained the area to be zoned was broadened by 1974 to include all public and private lands and all natural resources within the unorganized territories. The increasing scope of the Commission's mandate was tempered by a statutory scheme of zones that included a management zone defined as areas away from water and not currently developed for commercial or residential development essentially, the working woods where forest and agricultural management activities were allowed by right.

Administrative Structure of the Commission

The newly created Commission, with a staff of two, was located in the Governor's office and was not a separate department. The Commission currently has a staff of 27, located in five regional offices and one central office in Augusta. This was a common location for Quiet Revolution initiatives, as the administrative state was still young and appropriate departments to house these special area programs either did not exist or were just being formed. At the time of LURCs creation the only other relevant department would have been Forestry, whose watch over the North Woods at the time many included in the rationales for LURCs creation in the first place. The Department of Forestry was not structured as a regulatory agency, its mission by and large was to provide statewide silvicultural expertise and fire control to large and small landowners. So aside from not having a logical and willing department to incubate the Commission, housing LURC in the Governor's office also gave the nascent program a measure of political and budgetary cover.

LURC was one of the earliest Quiet Revolution agencies created in the State of Maine, but others followed quickly on its heels in the early 1970s. The creation of the Department of Environmental Protection (DEP) out of what had been the Environmental Improvement Commission created some jurisdictional questions that still challenge both entities to this day. One of Maine's signature environmental protection laws is the Site Location and Development Act ('Site Law') which, because of several proposals for oil terminals along the coast, for the first time created categories of development that were determined to be of state-wide significance that required state permits.[5] While the Site Law is not a planning and zoning statute in that it conferred no authority to create comprehensive plans or prospectively locate (zone) development on the landscape, it did create duplicative DEP and LURC permitting for large development in the unorganized territories. This duplication raised a great deal of hackles in the private sector as unnecessary, however early staffers have interpreted the legislative intent of the creation of the Site Law, which was also enacted in the 104th Legislature in 1969, as meaning to ultimately combine the two new environmental entities into one agency. As such, the duplication of permitting jurisdiction was by design in order that a merger would not require significant new enabling legislation (Haskell, 1973).

A merger never occurred due to a number of circumstances, but a principal one appears to be bureaucratic turf battles and administrative concerns how LURC, which is essentially just a state-appointed, local planning and zoning board, would function in what had become by 1971 a strictly regulatory and licensing agency with a mandate that included hazardous waste clean-up, air pollution control, as well as land-use permitting. The lack of a completely rational administrative reorganization following the early heady years of the administrative state was certainly not for want of trying, as many different proposals were and still are developed. One early one even proposed splitting the planning and permitting functions of the Commission into separate departments, where the State Planning Office would administer the Comprehensive Plan, and the DEP would issue permits and enforce rules (Haskell, 1973). If enacted, this proposal would have either made for some innovative cross-department collaboration or hopelessly fragmented the Commission's statutory mission. Given administrative realities fragmentation was a likely outcome.

This duplication of permitting authority due to the Site Law, and other statutes on protected natural resources existed between the two agencies until 2000, when the 119th Legislature removed most of the duplication by granting LURC the sole state authority to issue wetland permits in its jurisdiction, as well as stipulating that LURC would no longer need to issue a permit for many large projects requiring Site Law review. That the duplication endured for as long as it did fueled much of the perennial legislation that seeks to reorganize, restructure, or simply eliminate LURC.

Despite protests by some critics concerned that the store was being given away, the recent removal of duplication has had the effect of garnering significant legislative support for the Commission in general, and in particular when the Commission has sought to strengthen other elements of its statute. This was true most recently when the current 120th Legislature removed a problematic exemption in the Commission's subdivision definition, having declined to do so on two other occasions in certain measure because of the perception of excessive regulation and red tape. The streamlining of jurisdiction has allowed the Commission to significantly strengthen its authority in other areas not subject to either agency's jurisdiction.

Political Representation and the Unorganized Territory

The earliest incarnations of the Commission went through a number of significant changes in membership that foreshadow some of the continuing political debate about the adequacy of political representation for full-time residents of the unorganized territories. The lack of local government in an area such as New England, where annual town meeting is still the norm, precipitates surprisingly frequent rhetoric about the tyranny of State oversight.

The first Commission was structured to include state employees and the Commissioner of the Department of Conservation. The presence of state employees was quickly objected to by rural legislators because they felt that the unorganized territory's local planning board should be as representative of the jurisdiction as possible and state workers living hours from the jurisdiction did not have a personal stake in the rules they were to enforce.

Later in 1976, the Commissioner of the Department of Conservation was removed from LURC in order to free the board from direct political control by a Governor's appointee. So by the middle 1970s a formula was eventually settled upon where five seats were to be held by individuals with particular experience in the jurisdiction, including forestry, business and industry, and wildlife expertise. The two remaining seats did not have a particular constituency. The only concession to those wishing to have the Commission wholly comprised of full-time residents was a requirement that the Governor 'consider' persons living in, or near, the jurisdiction.

This representation scheme, which was designed to broker both the local and state interests present in the Commission's enabling statute, was never widely accepted and continues to be fought over in almost every legislative session since the late 1970s. Proposals are repeatedly introduced to either make the Commission elected, increase the number of jurisdiction residents, or to even turn the entire jurisdiction into one large municipality in order to assume 'local control.' The latest version of the compromise has two residents on the Commission and all of the members with some significant degree of experience in living or working in the jurisdiction. Such is the strength of local government and home rule in New England that its absence is seen as almost tyrannous. County government has little comparative authority and so does little to fill the ideological vacuum created by having only state representation in government for the unorganized territories. Conservative politicians and property-rights activists have long targeted LURC for special opprobrium. Where their assaults on the membership structure have been more or less successful, their repeated attempts to abolish the Commission entirely have rarely made it out of Legislative committees.

Where frontal assaults have failed, flanking maneuvers have beset the Commission's funding. With limited county government in New England, the State has assumed most of the tax and spend authority for the unorganized territories, including crafting budgets for schools, roads, land-use regulation, and fire protection. This State-level fiscal control is opposed by property rights activists alleging that they are modern-day sufferers of the revolutionary scourge of taxation without representation. This allegation came to its most controversial head in 1983 when for the first time almost the entirety of LURC's budget was to be paid by property taxes collected only from the unorganized territories. Previously the Commission's budget was taken from the State's General Fund, which is comprised of diverse revenue sources collected statewide. This attempt to require the property owners in the jurisdiction to shoulder the entire budget of the Commission where membership on the board and the enabling statute both had a significant statewide element was furiously protested and quickly changed. The final compromise position was that the property tax collected from the unorganized territories would fund 10 per cent of the Commission's entire budget, the remainder to be paid from the General Fund, and to the State as a whole.

Planning and Zoning Scheme Capability Planning

The Commission's zoning scheme, like the other regional plans developed in New Jersey's Pinelands, New York's Adirondack Park, and the Lake Tahoe Basin, was derived from capability planning methodologies. An early report drafted for the Commission outlined a methodology taken directly from the work of G. Angus Hills and Ian McHarg with a number of significant simplifications (Hills, 1961; McHarg, 1969). The Commission staff, like many others around the country at the time, was keenly interested in capturing the intricacies of ecological process and function in plans and rules. While the early 1970s was a time of widespread public interest in popular ecology and environmentalism, planners' interest in capability planning was the result of Ian MacHarg's captivating rhetoric in his influential *Design with Nature* published in 1969.

But unlike the Adirondack Park Agency and the ambitious leadership of its first planning director, the Commission staff quickly tempered their consideration of capability mapping because of the size of their jurisdiction and the limited amount of available information. It would ultimately take the Commission six years to formally adopt final zoning maps and rules for the entire 10.4 million acre jurisdiction. This contrasts markedly with the 18 months that the Adirondack Park

Agency required to create its final zoning map and land use plan for the 6 million acre park.

In early internal reports and staff dialogues that sought to establish the mapping and zoning methodology, Commission staff and analysts realized their limits of available information, 'therefore it would seem that the initial zoning should be more a function of location and demand than of gross ecological generalization' (Laws, 1971, p. 6). This translated into the understanding that LURC's 'land use zoning will have to be based on predominately cultural factors, but regulation can be based on information gathered by the landowner's making development proposals'. (Laws, 1971, p. 6).

Despite the potentially vast information requirements of capability mapping on millions of acres, staff determined that they should remain focused on protecting water quality because of the amount of substandard 'camp' development occurring along lakes and rivers. The impacts of most concern were eutrophication from faulty or excessive subsurface waste water disposal systems placed too close to water bodies, and excessive vegetative clearing for development or timber harvesting.

The initial zoning scheme was a hierarchy of districts (zones) that were discerned from both resource assessments and land use inventories. The process was to be based primarily on geography at the top of the hierarchy, where the jurisdiction would be first broken into four different physiographic regions coastal, mountainous, plain, and undulating. These regions would then be broken down into watersheds for all third-order streams or rivers. Then each watershed would be the elementary planning unit and could be ideally evaluated from an ecological process vantage point as additional zoning designations were made. However, when staff began delineating broad district boundaries early in the process, existing jurisdictional boundaries carried the day as townships, rather than third-order watershed basins, became the basic planning unit.[6] It was simply an easier methodology to follow the existing civil divisions when developing draft zoning maps and inventories.

After the basic mapping unit was determined, four districts were developed that broadly reflected the character of existing uses and development. Development, or areas of existing residential, commercial, or industrial use. Protection, that is water bodies, high peaks, wildlife habitat, wetlands, aquifer recharge areas, and unusual cultural features such as historic sites. Management or commercial timber harvesting and agriculture and related development such as mills, kilns, and landing yards. Holding or undeveloped areas adjacent to development districts. The four districts, each with its general purpose, would then each have a set of more specific subdistricts that would be the actual zones that specified allowed uses and development standards.

Development Subdistricts

Early methodologies proposed to filter development districts through a series of determinants that were principally visual and scenic. Areas of existing commercial and industrial development were to be zoned for such continued uses only where views were short and clearing was minimal. Residential areas could be zoned such where views were longer and distant and clearing had occurred or vegetative cover was less extensive (Laws, 1971).

These aesthetic determinants that included presumptions of extinguishing non-conforming uses, or mitigating existing clearing conditions were far in excess of the legislative intent of the LURC statute. The legislative debates on the LURC statute in 1969 focused closely on the issue of non-conforming uses and the fear by opponents that the new land use planning entity would seek to extinguish the many small and inexpensive camps that its champions felt were unsightly, visually intrusive, and environmentallydeleterious (Lacognata, 1974). As a result of this early debate, the Commission's statute clearly enumerates specific protections for non-conforming structures that allows them to remain, be rebuilt, and expand. The initial proposal for development subdistricts appears to run counter to the legislative intent on existing commercial and industrial development.

Ultimately, development districts were delineated only based on the amount and type of existing development occurringwithin a specific area and did not have criteria on views or clearing. For example, the residential subdistrict is laid down anywhere where there are four or more homes within a 500 foot radius. The subdistrict boundary is drawn relatively tightly around this node of existing development and includes little additional vacant land. There were only four original development subdistricts: residential, commercial/industrial, general development, in effect, rural villages and planned unit development. A few others have been created,but there are comparatively few development districts given the size and diversity of the jurisdiction's settlements.

Protection Subdistricts

Just as initial strategies for development subdistricts were more elaborate than what was finally used, areas around lakes were first proposed for assessment using distinct 'lake zoning determinants' gleaned largely from the work of G. Angus Hills in Ontario, Canada. These detailed determinants were developed so water bodies could be zoned to protect their natural functions and values, not necessarily their existing use.

These determinants sought to zone lakes both for surface water use as well as the type and density of shoreline development. The surface water use determinants were crafted to differentiate between types of motorized uses i.e., water-skiing, float plane landings, horsepower size of motors, non-motorized uses i.e., canoes, sailboats, and the types of activities, i.e., fishing, swimming. Initially, staff proposed two shoreline development determinantsto govern densityexisting building setbacks and subsurface wastewater treatment system guidelines. The wastewater treatment guidelines were to be in turn based on the physical characteristics of individual lakes. But in the final initial proposals, density was to be regulated by a ratio of shoreline to surface area, presumably where a larger ratio would allow more development. Initials proposals do not elaborate how this ratio was to relate to the lake zoning determinants.Finally the early zoning scheme proposed a separate category for those water bodies that were more than five miles from an existing road. These water bodies were to be classified as 'remote natural environment lakes' and would be subject to a ten horsepower limit on boats and development at no more than one dwelling per mile of shoreline.

The actual methodology for laying down protection districts, especially on lakes was much simpler than proposed and relied on existing information sources to delineate a range of important resources in addition to rivers and lakes, including deer wintering areas, aquifer recharge areas, wetlands, flood plains, and steep slopes. Detailed assessments of water bodies using the other proposed determinants were not done and only one subdistrict was created for all lakes not currently developed to the threshold of the residential development subdistrict. The finer scale resource assessments of the jurisdiction's many lakes anticipated in the staff's early proposal would happen later in the 1980s and be implemented as overlay zones.

Management Subdistricts

The original 1967 enabling legislation specified a fairly narrow jurisdiction for the then to be named Wildlands Use Regulation Commission that threaded along roads and around accessible water bodies to a depth of just 500 feet. All other land, which was presumably in active timber management, was not to be regulated. When the Commission's jurisdiction finally settled out to include the entirety of land in the unorganized territories, the idea of allowing forestry and agricultural management activities by right on the land outside of shoreland and other protected natural resources was carried forward as part of a political compromise with the timber industry. As such, the Management subdistricts were written as largely to excuse most timber and agricultural activities with the exception of larger commercial forestry facilities such as mills covering more than 5 acres. The Management district prohibited subdivision, but allowed residential dwelling units with a permit on any lot greater than the 40,000 square foot minimum. LURC does not utilize use zoning on timber lands where lot sizes for residential dwelling units were required to be as large as commercially viable small woodlot such as in the states of Oregon and Washington.

Where the zoning scheme largely exempted the Management districts, Maine does have a use-valuation program for timber lands, the Tree Growth Tax Program. This program was enacted in 1971 as an incentive to keep lands in commercial production by lowering property taxes to a state-determined value on each of several forest type stands. Penalty provisions exist for the removal of land from the program for development. The Tree Growth Tax Program has been both successful based on the amount of forestland enrolled, and controversial as a result of impacts on local tax base, and more recently whether recipients of the reduced taxes conferred by this program should be prohibited from charging any access or gate fees to the public wishing to use their land.

There are three Management subdistricts: General Management, Highly Productive Management, and Natural Character Management. But General Management is the only subdistrict that has ever been applied to a zoning map. This subdistrict governs almost 80 per cent of the acreage in the jurisdiction, or essentially everything else after development and protection subdistricts were delineated. The Highly Productive subdistrict is designed to include areas of prime or unique agricultural and forest lands, but has never been mapped due to what was at the time, a lack of detailed soils information for many areas of the jurisdiction. Landowners are also skeptical of the use of this subdistrict fearing that such a zone unnecessarily limits their options and is not needed to protect forestry and agriculture. The district is not particularly restrictive in terms of uses as residential dwellings on one-acre lots are allowed by permit.[7]

The Management Natural Character subdistrict is not a viable category because of its restrictivness. It proposes a limited range of uses for presumably wide areas of the jurisdiction's interior under active timber management. Aside from commercial forestry only limited types of primitive seasonal recreational dwellings are permitted uses in the Natural Character subdistrict, in order to maintain the 'natural outdoor flavor and spirit of certain large undeveloped areas' of the truly remote and largely unpopulated areas of the jurisdiction.[8] This subdistrict comes the closest to use-zoning and as such was strongly resisted at the first public hearings in 1971 on the proposed implementing regulations. As a result of the negative comments, this subdistrict was amended so that it could only be applied with the landowner's consent, one of the two subdistricts to include this provision. As well, language was removed from the description and purpose statements that indicated the use of this subdistrict was designed to be 'an alternative to federal ownership'.

The last two words are fighting words in Maine, which has only about 200,000 acres of land in federal ownership, most of which is national forest or wildlife refuges, a small proportion even by Eastern United States' standards. Discussions about a national park in the Great North Woods have come and gone since the 1920s, with all proposals avidly fought by a wide cross-section of Mainers. Indeed, Governor Percival Baxter's acclaimed gift of the 200,000 acre Baxter State Park in the 1930s to the people of Maine was an attempt to prempt the creation of a national park.[9]

Although never mapped, the Management Natural Character subdistrict surfaced most recently in legislation proposed in 1997 by a coalition of conservation organizations, would have required the Commission to rezone several million acres to this subdistrict to control what was seen as an unacceptable amount of recreational second home development allowed in the first twenty years of the Commission's existence. This proposal was quickly dubbed the 'wilderness zoning' bill and was defeated by the full Legislature after an acrimonious debate.

Since the legislative fight over down-zoning the interior of LURC jurisdiction, several large-scaleconservation easements have been sold, triggering an unprecedented amount of land conservation. To date, two owners have sold development rights on almost 800,000 acres and placed the land under conservation easements that extinguish any further recreational development while providing for continued timber management. In addition, a private non-profit land trust has purchased almost 150,000 acres from timber companies surrounding the St. John River along the Quebec border. These publicly and privately financed deals have literally triggered a deluge of offers from many other large landowners interested in severing the development potential on their holdings while there is ample public and private money to purchase development rights on forestry land. At this writing, there are another two million acres of land under option for conservation, and the rumors are that there will continue to be additional million-acreproposals announced for the next several years. It is likely that close to 1/3 of the Commission's jurisdiction will be under easement restricting development within the next two years. Currently 1.4 million acres are already under easement or the subject of active fundraising. This is a wholly unexpected course of events for the Commission that has the practical effect of resolving without regulations – a trend of inappropriate development that it outlined in its 1997 revision to its comprehensive plan.

Holding Subdistricts

The final original zone was the Holding district, defined as reserve areas adjacent to development districts that were to accommodate growth once the development districts were, in the words of the initial regulations, 'saturated'. The Holding district was eliminated in 1973 in the first revisions to the implementing regulations. The staff did not feel adequately able to develop a methodology to size these districts given the uncertain demand for development across the disparate regions of the Commission's very rural jurisdiction. The demise of the Holding district cast the balance of the zoning scheme as essentially spot zoning where all designated development subdistricts were, by definition, built out and proposals for new development almost always required land in Management or Protection subdistricts to be rezoned to a development subdistrict in order to allow the proposed use. The Commission has labored under the inadequacies of this reactive zoning scheme throughout its existence. While spot zoning relieved the Commission of the work required to craft appropriate areas for new development, the end result has been more protracted struggles on the proposals submitted by developers and less control on the location of new development.

Adjacency Criteria as Central Policy

As in other special area programs, recreational second home development is the most prevalent and significant land use being regulated in LURC's jurisdiction. There has not been a significant amount of large-scale recreational development such as ski-areas, amusement parks, or destination resorts. Maine, long named 'Vacationland' on state license plates, has been a highly desirable location for both residents and non-residents to buy second homes created through comparatively small-scale subdivisions. At least since the 1940s, among the fifty states Maine has had the second highest number of vacation homes as a percentage of total housing stock where 15 per cent of homes were camps, cottages, and cabins.

In order to protect what it sees as the four principal values of the jurisdiction – the economic value of the timber lands, diverse and abundant primitive recreation opportunities, diverse and abundant high value natural resources, and undeveloped natural character – it has been one of the Commission's central goals since its inception to direct this type of camp development to appropriate areas of the jurisdiction. In practice, this has meant areas already developed, or to locations in the fringe of the jurisdiction in order to leave the interior or core dedicated to working forest and remote undeveloped lakes. This goal is implemented by the application of the Commission's subdistricts in conjunction with its zoning test of 'adjacency'.

As a result, subdivisions are not an allowed use in any of the Management or Protection districts, which together cover approximately 95 per cent of the acreage of the jurisdiction. As such, for a subdivision to be approved in these areas, the property must be rezoned to a Development subdistrict. Aside from the typical statutory approval criteria, a rezoning for a subdivision must meet the adjacency threshold, which requires any new proposal to be within one road mile of either another Development subdistrict or a 'similar pattern of development'. The adjacency test is extremely simple and is intended to keep new development moving in coordination with existing development, avoid leapfrogging, and maintain a very lightly settled interior. This locational standard is a fairly uncommon strategy for controlling development, where most special area programs either implement strict use zoning, such as in Oregon and Washington with forest use zoning, or include large blocks of protected public land as in the Pinelands and the Adirondacks.

The prohibition on subdivisions in most of the jurisdiction and the requirement for rezonings to meet a locational threshold would seem to be an effective tool for controlling the pattern of development. It has however been undone by the statutory definition of subdivision, which until very recently contained substantial loopholes that allowed for the creation of many lots outside of the adjacency standard. The most prominent of the loopholes was the exemption for lots in excess of 40 acres. Originally enacted to allow for the creation of small woodlots that would be managed for firewood or timber harvesting, this exemption has been used to parcelize almost 200,000 acres in both Management and Protection subdistricts into what has become not small woodlots but recreational camp lots whose owners do not manage for timber.[10] The frequent use of the 40-acre exemption has had the effect of scattering development throughout both the fringe and interior of the jurisdiction as well as negatively impacting the commercial forestry land base by removing excess land from production (Land & Water Associates, 1994a). One simple measure that indicates the change in development patterns brought about in part by these loopholes is the decreasing numbers of unpopulated townships, whose numbers have dropped from 176 in 1970 to less than 109 by 1991, a decline of 40 per cent (Land & Water Associates, 1994b).

The Commission has tried to eliminate the 40-acre exemption on several occasions in the last 15 years, initially succeeding in 1989 and 1991 by restricting the rate of their creation and limiting them to areas away from water bodies. In 2001, the State Legislature, at the Commission's request, has now completely closed the loophole by applying the exemption to only those lots transferred for forestry and agricultural management. All other types of structural development would require subdivision review and approval. This last statutory change should enable the Commission to fully implement its central policy of ensuring that development follows existing patterns. However, it remains to be seen as to how the significant amount of existing lots created using the past exemptions will act to create a pattern of development sufficient to meet the adjacency test for new subdivisions seeking Commission approval. In short, despite the closure of loopholes the die may already be cast.

Lake Management Classifications

In 1977, shortly after the final regulatory framework and zoning maps were adopted throughout the entire jurisdiction, the Commission and other State agencies began a comprehensive assessment of the almost 3,000 lakes in its jurisdiction that would produce a very similar regulatory scheme to that first proposed for water bodies during the original zoning in the early 1970s. Beginning in the middle 1980s assessments of each lake's resource characteristics, including fisheries, wildlife habitats, scenic resources, shoreline configuration, cultural and historic resources, and physical characteristics were tabulated over a period of several years. By 1990, the Wildlands Lake Assessment produced a rating scheme that identified a hierarchy of lakes that began with those that received at least two 'outstanding ratings' making them lakes of statewide significance.[11] The rating scheme ended with lakes of moderate value deemed significant at a local or regional level. These generalized resource ratings were used by the Commission to produce a management classification system that produced what was in effect an overlay zone to some 400 lakes.

There are six management classifications which govern lakes from those 177 'remote ponds' that support unique cold-water fisheries where further development is precluded and have no vehicular access within a one half mile of the lake to 'high value, developed lakes' where any further subdivision is required to be clustered in order to preserve some semblance of undeveloped shoreline.

While the management classes have been successfully adopted and have proved to be a very significant and important element in the Commission's regulatory framework, they are not comprehensive or yet fully implemented. While every lake surveyed during the resource assessment stage was given a resource rating, not every lake in turn has received a management classification because of the difficulty in crafting a reasonable number of management classes for the wide range of lake types and development circumstances in the jurisdiction. In fact, numerically the vast majority of the lakes do not have any management classification, but are essentially categorized as 'all other'. These 2,600 other lakes contain both high and low value water bodies that are only assessed on a project by project basis, which is not unlike the final outcome of the initial planning and zoning scheme for the dry land, which created what looks a lot like spot zoning operating under some broad locational and environmental impact policies.

The management classificationsystem is not yet fully implemented. For instance, Management Class three, which is one of the more prospective classifications and is applied to 36 undeveloped lakes in the fringe of the jurisdiction, has not been fully elaborated in terms of specific developmentdensity and standards.[12] Large landowners who participated in the management classification process have consistently complained to the Commission that while strict standards were applied to three of the six classifications, including prohibitions on development or a density threshold of one seasonal dwelling per mile of shoreline, the Management Class three, which deemed lakes to be 'potentially suitable for development' did not have criteria that clearly stipulates the extent of allowed development. While there are not hard numerical standards for this lake class, as exist in the other classifications, the Commission removed the requirementthat development proposed for these lakes meet the adjacency standard, which is a significant threshold criteria for approval of new development.

As well, and what makes this classificationthe closest to a capability designation, each of these lakes was hypothetically subjected to a maximum shoreline build out, using minimum dimensional requirements, in a phosphorous allocation model that predicts water quality impacts. For a lake to meet the Management Class three criteria of potentially suitable for development, maximum shoreline build out could not increase the phosphorous concentration by more than one part per billion (Dennis, et.al, 1983). This type of capability planning is far less technically rigorous than that applied to Lake Tahoe, but a significant step ahead of the Adirondack Park Agency, whose staff have lamented the lack of consideration that was given to lake capacity in the creation of its land use plan and zones.

Nonetheless the landowners' concerns about the work still to be done on the Management Class 3 lakes are reasonable. In response, the Commission has created and adopted specific standards for two of the larger of these lakes to serve as a potential model. Development on these two lakes must meet an average density of one dwelling unit for every 400 feet of shore frontage and have a minimum of fifty per cent of the shoreline permanently protected from development.

State and Municipal Relations

The state and local relations within the jurisdiction of the Commission are very different from other Quiet Revolution regional entities, including the Adirondack Park Agency, the Pinelands Commission, and the Lake Tahoe Regional Commission. A central tenet of these campaigns was the inadequacy of local control, or home-rule, where municipalities had historically retained the authority to control most, if not all, land use decisions. Regional and state-planning schemes were enacted in the late 1960s and early 1970s throughout the United States in order to overcome what was seen as the problems of parochial and local decision-making. Broader mandates and wider perspectives were needed to reverse the haphazard development and environmental degradation of the status quo. On most other fronts of the Quiet Revolution, the regional and statewide schemes operate differently than in Maine. Elsewhere the standard model was to remove most, if not all, local land-use authority from municipalities and vest it in these new administrative agencies. Land use authority would remain with the agencies until localities adopted planning and zoning schemes consistent with the regional ones. As described in the other chapters, there have been varying degrees of success with this model with few towns rising to the mandate in the Adirondacks, but all those in the Pinelands doing so as a result of strong fiscal incentives.

In Maine, the model was different because there was (is) no local government within most of the area of the Commission's jurisdiction, so there was no local authority to remove. Much of the unorganized territory, which is officially designated as townships in Maine law, remains not only as sparsely populated as it was in the 18th and 19th centuries but still retains the original survey designations which serve as town names. Most townships are still referred to by the township and range designations used by the original colonial-era surveyors. For someone new to the state (or 'from away') the names T6 R14 WELS or T3 R7 BKP do not immediately trip off the tongue.[13]

Municipal law in Maine provides for four categories of municipality - city, town, township, and plantation. There are no cities, generally administered by a professional town manager and board of selectmen, within LURC jurisdiction. There are seven towns, which by law have the authority to regulate land use, that have elected to have the Commission function as their planning and zoning authority. The vast majority of the entities, some 411, are townships, which have no local government or local control at all. As well, some have no residents either, just trees. The remaining civil division - plantation - is a curious relic of the colonial settlement of New England. Plantations are, no offense intended, 'sort-of' towns. They have the authority to hold a town meeting and elect a board of assessors that administer the tax and spend authority. However plantations do not have land use authority. The 32 plantations that range from the famous Monhegan Island off the coast to Sandy River in the Western mountains, to Cyr in northern Aroostook County, are all under the Commission's jurisdiction.

The Commission was in effect the first effort to enact planning and zoning in the unorganized territories. However there are options for a township or other municipality in LURC jurisdiction to assume local control of all land-use decisions governed by the Commission. The Commission's enabling statute restates existing provisions of municipal law that allows for unorganized territories at their discretion to organize and assume town status. What the LURC statute added were provisions that any township seeking to organize must develop a local plan and zoning ordinance that is 'not less protective of the existing natural, recreational, or historic resources' governed by the Commission's plan and zoning ordinance. Prior to the formation of the Commission, no township seeking to organize as a town was required to develop a plan and zoning ordinance, as this is a discretionary authority of towns in Maine.

The opposite course of a town deorganizing and so shedding local control in order to become a township is also prescribed in Maine statute. However, on balance most movement has been from townships organizing out of the Commission's jurisdiction. Since 1971, ten townships or plantations have assumed local control and left the Commission's jurisdiction, only three have entered. Generally the towns that leave LURC jurisdiction do so to remove State control over land-use decisions. The motivation for towns entering LURC jurisdiction are more ambivalent, where fiscal concerns generally push residents to give up local control. The latest town to deorganize Madrid in the western mountains gave up local control in 1999 in order to control taxes. The highly chronicled deorganization of this small town (population 198) was precipitated by the enrollment of 80 per cent of the town's acreage in the Tree Growth Tax program. The Town's finances suffered in part because the State did not fully fund the reimbursement element of the Tree Growth Program through most of the 1990s, where local coffers are required by statute to receive 90 per cent of the lost property tax revenues. In fact, payments did not again reach the 90 per cent level until 1997.

In addition to the Tree Growth impact, rising school costs were seen by local residents as increasingly difficult to meet on a wholly residential tax base. Almost 70 per cent of the town's taxes went to provide education for its 36 students. As one proponent of deorganization noted, 'I've been chairman of the School Committee for 20 years, and I always feared a family moving into town with several children because of the increase in taxes' (Waterhouse, 1999). In the end, after receiving approval from the State Legislature to put the question of deorganization to the residents, the Town voted 76 – 38, exactly the two-thirds required to pass the initiative. The closeness of the vote reflected the ambivalence felt in the town about dissolving the 160-year-old municipality.

The drive to reduce local tax burdens by deorganizing - and so spreading school, road, and other infrastructure costs across the other ten million acres of unorganized territory - is not necessarily a sure bet. While the tax burden is generally significantly lower in the unorganized territory, state tax officials were constantly careful to remind residents of the prospective township that their tax rates might even increase after a revaluation was conducted upon deorganization. In the end, the gamble paid off, with Madrid's tax rate declining by 43 per cent after deorganization.

While deorganizing is a gamble, it is clear from informal discussions around the jurisdiction and in the Statehouse that there are numerous other towns and plantations that are seriously discussing the move as school expenses rise against an increasingly smaller population and a disproportionately residential property tax base. Were municipal deorganization to pick up, it would buck a long-standing trend where most efforts fail at popular votes of residents. Only three out of nine initiatives in the last twenty years have gone to completion. As one resident of a town that declined to deorganize noted, 'Some of us have been in this town for generations. We're hard-working people who've cared for each other for a long time. I'd like lower taxes, but at what price?' (Clancy, 1998).

This common sentiment among those declining deorganization shows that these residents abide a very strong connection between town government and community. They feel strongly that when there is no longer elected town officials or a town clerk, when roads are maintained by the county, and schools administered in the State capital, then there is no longer a sense of community. It is debatable that the elimination of New England town meeting and some elected positions has an impact on the many other connections, relationships, and interdependencies of small town life. However, it is clear that the association of government and community is visceral. As Madrid selectman Melvyn Webber, a vocal opponent of the town 'going out' felt the town has lost 'everything beyond the pocketbook' and the gift of living someplace 'where you can stand up in a town meeting, and say what you want, and be heard ...being a town MEANS something' (Zimet, 1999).

What is probably more apt is that deorganization is one of many symptoms of the broader changes in these rural communities. The town of Madrid first lost its industrial and manufacturing base in the early twentieth century, after which the town's infrastructure began to fade with the Grange hall and post office having burned down, the local store closed, and the only dance hall in town standing empty most of the year.

Tribal and State Relations

The small towns are not alone in feeling the tension between their sense of community and their relations with the State. Maine's Native American populations struggle as well. As a result of a lawsuit filed by the Passamaquoddy and Penobscot Tribes that laid claim to two-thirds of the land area of Maine, the Federal government signed a settlement in 1980 awarding the Tribes each $40 million dollars and the authority to purchase land that would be held in trust by the Federal government as commonly owned tribal property. The settlement extinguished the tribes' land claim, based on the argument that their colonial land treaties signed were not valid because they were never ratified, and negotiated their political status in Maine as federally recognized tribes.

The compromise enacted following the lawsuit remains contentious largely because it created a unique jurisdictional status for the two tribes.[14] State officials were keenly concerned that the tribes not have the same sovereignty as exists for all other federally recognized tribes in order to prevent what then Governor James Longley called 'a nation within a nation'. As such, the Maine Indian Claims Settlement Act categorized the Indian trust lands to be purchased and the existing reservations as municipalities subject to State laws except where the Settlement enumerated otherwise. The municipal status has pitted the Tribes against the State in many areas, not the least of which is land use regulation. The State has maintained the position that the Tribe's land use decisions – including the controversial authority to build a high-stakes bingo facility – are subject to the Commission's jurisdiction unless the Tribe, like other municipalities, goes through the process of formally organizing and adopting a land use plan and implementing ordinance approved by the Commission. The Tribes bitterly resent their circumscribed authority, and as Passamaquoddy tribal leader John Stevens noted, 'If I had it to do all over again, I would have gone to court. To me it's a sad day when I instructed my people to compromise' (Suggs, 2000).

The jurisdictional dispute is unlikely to be resolved any time soon, despite the repeated efforts to formally remove Indian trust lands from LURC's jurisdiction. However it appears in recent years there has been a softening of the State's position on land use authority as it gains more experience with the Tribes administrative and civil capacity. Though the State may be compromising on land use, it has not conceded to Tribal authority on water pollution permitting. The Tribes and a number of paper mills (whose position the State supports) remain in litigation with little opportunity of a compromise solution possible.

Conclusion

While many of the issues that confronted the jurisdiction in the early 1970s remain the same, Maine is no longer the place it was almost thirty years ago. Structural changes in the state, nation, and global economy will continue to influence not only business, but recreation in the Great North Woods. As a result, the Commission's planning and zoning scheme must continue to adapt. Indeed this is the singular challenge of all administrative agencies in the modern liberal state. Circumstances, constraints, and opportunities arise often much quicker than public entities can respond and still govern equitably. It is the responsibility of the Commission to effectively deliberate on the changes in land ownership, development patterns, and resource demands occurring throughout the jurisdiction. If this requires fundamental change in the underlying zoning scheme or operative policies on rezoning and adjacency, then the Commission and its staff should be up to the task of changing course. But while the current scheme shares many of the admirable traits of the New England Yankee in that it is sturdy, understated, and hardworking, it would be disappointing if the Commission also abides that other Yankee inclination to do things the way they've always been done. That would keep a big place and its big plan from getting better.

Notes

1 Maine's ranking is currently 37th, where it has remained for almost a decade and at $24,603 is a full $10,000 behind the other New England states. U.S. Bureau of Economic Analysis, October 2000.

2 The ten firms are: J.D. Irving, LLC (1.5 million acres), International Paper (1.4 million acres, Seven Islands Land Company/Pingree Family (960,000 million acres), Prentiss & Carlisle (935,000 acres), Plum Creek Timber Co. (912,000 acres), Mead Corporation (661,000 acres), McDonald Investments, Inc. (656,000 acres), Great Northern Paper Co. (380,000 acres), Hancock Timber Resources (343,000 acres), and Huber Resources Corp. (300,000 acres)

3 The word on the real estate street is that these very large recreational lots with miles of shorefront on pristine lakes are now of such a scale that the moniker 'kingdom lot' is no longer apropos. The latest sales lingo has these listed portentously as 'empire lots'. Apparently the moguls are flocking to Maine as the real estate is cheaper than the over-heated market for ranches in Wyoming and Montana.

4 McKee was a history professor at Maine's Bowdoin College, whose tactics were very much like Verplanck Colvin's 19th century photography of despoliation in the Adirondacks, publicized the cause of unplanned land development.

5 Title 38, Section 481-490, Maine Revised Statutes Annotated.

6 The absence of a watershed based zoning scheme was also a complaint leveled against the Adirondack Park's Private Land Use and Development Plan.

7 A very common local bromide is that you can't stop a tree from growing anywhere in Maine.

8 Section 10.15, C, 1 of the Commission's Land Use Districts and Standards.

9 Governor Baxter did not much favor the State's stewardship either, so he created a quasi-independent authority to run the Park according to his guidelines and principles with an endowment he donated as well. An aspiring heir to Baxter's legacy, Roxanne Quimby, sole owner of a profitable natural beauty products business has purchased 8,500 acres of land over the last two years in the North Woods with the express purpose of donating it to the National Park when she believes it becomes a reality. She does not share Baxter's distrust of the Federal Park Service.

10 Management objectives of these properties can be inferred from the percentage of enrollment of these 40-acre parcels in the state's Tree Growth Tax Program, which assesses land at its value for timber production and not its highest and best use. On average less than 50 per cent of the several thousand parcels in the 40-60 acre size class are enrolled in this program, and a survey conducted by the Commission in 2000 of these lot owners revealed that 44 per cent of these owners indicated they managed their property for timber.

11 Rankings of outstanding, significant, or no rating were applied based on numerical scoring rubrics of a number of categories within each resource being assessed.

12 "Fringe" is defined as within two townships of the jurisdiction's border. Land beyond the fringe is categorized as "interior" as seen as been less appropriate for extensive development in the Commission's policy statements and Comprehensive Plan.

13 T6 R14 WELS is the sixth township, which runs west to east, in the 14th range, which run south to north, measured west of the easternmost line of the state (WELS), which is the border with the Canadian Province of New Brunswick. Others include "BKP" for Bingham's Kennebec Purchase, "IP" for Indian Purchase, et al. The survey designations are well-liked and have developed a long tradition of usage (even where some townships have adopted alphabetic names that have gained popular currency) that continue to resist attempts at change. The most recent effort arose as a result of

statewide efforts to implement an emergency response program using police dispatchers and a uniform statewide addressing system integrated into a geographic information system. Developers of the system proposed to eliminate the township and range nomenclature as they felt it was easy to transpose numbers and so misidentify townships. Their proposals to implement an alphabetical naming system have been repeatedly defeated in the Legislature.

14 The compromise was codified as the Maine Indian Claims Settlement Act, Title 30, Section 6201, et seq., Maine Revised Statutes Annotated.

References

Austin, Phyllis (2001), 'Evicted: Northern Maine Camp Lease Threatened by New Owners', *Maine Times*, May 24-30, 2001, pp. 4-6.

Bosselman, F. and Callies, D. (1972), *The Quiet Revolution in Land Use Control,* Council on Environmental Quality, Washington, DC.

Clancy, Mary Ann (1998), 'Whitneyville Residents Vote to Abandon Deorganization', *Bangor Daily News*, April 29, 1998, B1

Dennis, Jeff, et al. (1983), 'Phosphorous Control in Lake Watersheds: A Technical Guide to Evaluating New Development', Department of Environmental Protection, Augusta, ME.

Glidden, William T., et al. (1994), 'Integrating Land Use and Natural Resource Management: Final Report of the Land Use Regulatory Reform Committee', Office of Policy and Legal Analysis, Augusta, ME.

Haskell, James S., Jr. (1973), 'Administrative Review, LURC: Interdepartmental Cooperation and Reorganization of State Government, March 30, 1973', Typescript, Department of Conservation, LURC, Augusta, ME.

Hills, G. A. (1961), 'The Ecological Basis for Land Use Planning', Research Report No. 46, Ontario Department of Lands and Forests, Research Branch, Ottawa, Ontario.

Lacognata, Esther (1974), 'A Legislative History and Analysis of the Land Use Regulation Law in Maine', Typescript, Department of Conservation, LURC, Augusta, ME.

Land Use Regulation Commission (1972), 'Standards for Interim Land Use District Boundaries', Typescript, Department of Conservation, LURC, Augusta, ME.

Land Use Regulation Commission (1997), 'Comprehensive Land Use Plan', Department of Conservation, LURC, Augusta, ME.

Land and Water Associates (1993a), 'Trends in New Residential Development in Maine's Unorganized Areas – Amount and Location of New Residences By Region and Minor Civil Division', Typescript, Department of Conservation, MLURC, Augusta, ME.

Land and Water Associates (1993b), 'Location of New Development in the Wildlands: Where Residential, Commercial and Industrial Development Has Occurred in Relationship to Roads, Waterbodies, and Mountains', Typescript, Department of Conservation, LURC, Augusta, ME.

Land and Water Associates, (1994a), 'Impacts of Changes in Landownership Patterns and Development in the LURC Jurisdiction on Timber Production and the State's Forest Manufacturing Economy', Typescript, Department of Conservation, LURC, Augusta, ME.

Land and Water Associates (1994b), 'A Summary of the Commission's Current Land Use Policies and their Net Effects After 20 Years of Development in Maine's Unorganized Areas', Typescript, Department of Conservation, LURC, Augusta, ME.

Laws, C. (1971), 'A Systematic Approach to Land Use Zoning in Maine's Unorganized Townships', Typescript, Department of Conservation, LURC, Augusta, ME.

Legislative Research Committee. (1969), 'Report on the Wildlands Use Regulation to the 104th Legislature', Publication 104-1, Maine State Library, Augusta, ME.
Lewis, David J. (2001), 'Easements and Conservation Policy in the North Maine Woods', *Maine Policy Review*, Winter 2001, pp. 24-38.
Maine State Museum (1908), 'History of the Wildlands of Maine', Typescript, Augusta, ME.
McHarg, Ian (1969), *Design with Nature*, American Museum of Natural History, Garden City, NJ.
McKee, John (1969), 'Report on the Wildlands of Maine: Appendix F to Legislative Research Committee Publication 104-1, the report of the Subcommittee on Wildlands Use Regulation', Publication 104-2, Legislative Research Committee, Augusta, ME.
Scruggs, Roberta (2000), 'An Unsettled Settlement', *Portland Press Herald*, August 20, 2000, p. A1.
Turkel, Tux (2001), 'Kingdoms for Sale', *Portland Press Herald*, June 10, 2001, p. A1.
Waterhouse, Don (1999), 'Fiscal Trouble Kills Madrid', *Kennebec Journal*, November 7, 1999, p. B1.
Young, Susan (2000), 'Protecting Maine's Forests: Mainer's Debate Wood's Rules', *Bangor Daily News*, October 14-15, 2000, p. A1.
Zimet, Abby (1999), 'No More a Town', *Maine Sunday Telegram*, Nov 21, 1999, p. A1.

Chapter 8

A History of Planning in the Adirondack Park: The Enduring Conflict

Glenn R. Harris and Michael G. Jarvis

This chapter is divided into five sections. The first section outlines the significance of the Adirondack region. It describes the rationale for establishing both a New York State Forest Preserve and an Adirondack Park in the final decades of the 1800s. It also describes the important reasons for enacting both a State Land Master Plan and a private Land Use and Development Plan in the 1970s. The second section details events surrounding the implementation of these planning activities. A key point in these first two sections is that the Adirondack Park included a large proportion of private land when it was created in 1892. Although park proponents had hoped that the state would eventually acquire private lands within park boundaries, development pressure in the second half of the twentieth century necessitated, instead, a shift to a regulatory strategy in order to protect open space. The Land Use and Development Plan, enacted in 1973, governs private land through police powers administered by a state-level Adirondack Park Agency.

The third section of this chapter examines approaches for defining the Adirondack Park. Taking a cultural approach, this section outlines how permanent residents have perceived the Adirondack landscape differently from framers of legislative control. People on private lands within the park have needs that are not the same as those who formulated legislation, starting with concerns that led to formation of the Forest Preserve in 1885. Since then, differing constructions of the physical landscape have produced disagreements over policies of land-use planning. The conflict has been enduring.

The fourth section sketches some forces of change that have affected the Adirondack Park over the past thirty years. The fifth and final section documents how various organizations have responded to these changes. The organizations are not of one mind. Consequently, they do not speak with once voice. Some continue to advocate acquisition of land to add to the Forest Preserve; others emphasize the need for better management of state lands to accommodate burgeoning recreational demand. Some fear residential growth on private land and the threat of development to the open space character of the park; others worry, instead, about the effect of invasive species on ecological relationships within the natural environment of the park. Still others remain concerned about a declining tax base, stagnating employment, and the impact of socio-economic conditions on provision of public services. Despite these differences, all agree that planning is essential to minimize conflict and achieve respective goals. In fact, collective planning efforts have produced a certain tolerance, if not appreciation, for the concerns of others. Everyone concurs that additional resources are necessary to continue the planning process.

Historical Significance of the Adirondacks

Throughout the second half of the nineteenth century, the Adirondack region was thought to be significant because it was the headwaters for watersheds critical to the economy of New York. A Forest Preserve, consisting of widely scattered parcels of state-owned land, was established to protect these watersheds. When the Adirondack Park was established toward the close of the century, the land within its boundary included a mix of both state land, protected forever wild as Forest Preserve, and private land, which might eventually be acquired by the state. Increasingly, the recreational potential and mental health benefits of Adirondack forests were recognized. During the second half of the twentieth century, extensive natural areas were also looked upon as unusual biological habitat. Nonetheless, much of the land within park borders remained private, and a growing number of second-homes were beginning to dot the landscape. Then, about 30 years ago, these private lands were placed under stringent land-use controls in an unprecedented exercise of regulatory activity by the state of New York. The significance of the Adirondack region lies in its history, which gave rise to a park containing both public and private lands, and in its present, which represents a challenging exercise of land-use planning.

George Perkins Marsh nicely summarized the importance of Adirondack watersheds as early as 1864 in what would turn out to be one of the first comprehensive summaries of ecological principles, *Man and Nature*. Marsh, who grew up in Southern Vermont, had a varied career as a businessman, lawyer, representative to Congress, and ambassador to several countries. Influenced by principles of scientific resource management during his travels in Europe, Marsh said of the Adirondacks:

> It is evidently a matter of the utmost importance that the public, and especially land owners, be roused to a sense of the dangers to which the indiscriminate clearing of the woods may expose not only future generations, but the very soil itself...The State of New York, for example, has, in its northwestern counties, a vast extent of territory in which the lumberman has only here and there established his camp, and where the forest, though interspersed with permanent settlements, robbed of some of its finest pine groves, and ravaged by devastating fires, still covers the largest proportion of the surface...The felling of the Adirondack woods would ultimately involve for Northern and Central New York consequences similar to those which have resulted from the

> laying bare of the southern and western declivities of the French Alps...they have rapid slopes and loose and friable soils enough to render widespread desolation certain, if the further destruction of the woods is not soon arrested. The effects of clearing are already perceptible in the comparatively unviolated region of which I am speaking. The rivers which rise in it flow with diminished currents in dry seasons, and with augmented volumes of water after heavy rains. They bring down much larger quantities of sediment, and the increasing obstructions to the navigation of the Hudson, which are extending themselves down the channel in proportion as the fields are encroaching upon the forest, give good grounds for the fear of serious injury to the commerce of the important towns on the upper waters of that river...(Marsh, 1864, pp. 203-05).

This argument was championed on behalf on the Adirondacks by Verplanck Colvin, who explored and mapped the Adirondack region throughout the 1870s. The son of a lawyer and New York legislator, Colvin was appointed by the legislature in 1872 to be superintendent of a topographical survey of the Adirondacks. Using the rationale articulated by Marsh, he proposed in numerous reports that the Adirondack region be designated a park. Noting the importance of forests for protecting deep layers of peat moss as well as shallow layers of fragile soil, Colvin used the term 'hanging lake' to describe the substrate that could be destroyed by excessive lumbering (Colvin, 1871, p. 179).

In 1872, the same year that Colvin was appointed superintendent of the Adirondack topographic survey, a Commission of State Parks was created by the New York State Legislature. The function of the commission was 'to inquire into the expediency of providing for vesting in the State the title to the timbered regions lying within the counties of Lewis, Essex, Clinton, Franklin, St. Lawrence, Herkimer, and Hamilton, and converting the same into a public park'. From the very beginning, the commissioners emphasized the importance of this immense watershed to both the Erie Canal and the state's mills and factories. The commissioners warned that if the Erie Canal was to dry up due to excess destruction of the forested lands of the Adirondacks, farmers would be victimized by powerful railroad monopolies. If the waterways became impassable for trade, the railroads would be able to raise shipping rates as it wished (Graham, 1978, p. 76; Nash, 1982, p. 119).

Although the New York State Legislature ignored the commission's report, Colvin's idea of the forest as contributor to a stable water supply had taken hold among the public. The 'hanging lake' became the 'hanging sponge', composed of mosses and now dead leaves as well as other natural debris. Without the spongy forest floor, water would simply rush unchecked across denuded land and descend in damaging floods over the lowlands; none would be withheld to maintain the flow of rivers during dry seasons. Moreover, rushing water would sweep topsoil with it, gullying the land. Soils would deposit as sediment, clogging downstream rivers. Finally, there would be insufficient soil for vegetation to re-establish itself in the eroded mountains (Graham, 1978, p. 90). In summary, it was believed that without protection of the Adirondack forests, periodic droughts could render the state waterways impassable, and at other times, disastrous floods might overwhelm the lowlands. Residents of New York City became particularly roused at the lumber and mining companies alleged to be stripping the Adirondack forests.

The recreational importance of the Adirondacks has a longer history than the watershed arguments, even though the latter were ultimately more significant in the initial establishment of the Forest Preserve by the New York State Legislature. Vacationers and sportsmen visited the Adirondacks as soon as they were able to find their way to and through the forested mountains. Many of these visitors wrote books documenting their travels and extolling the recreational potential of the region (e.g. Headley, 1849; Hammond, 1857). None, however, was as insistent about the benefits of an Adirondack vacation as was Reverend William H. H. Murray, a Congregationalist minister from Boston, whose 1869 book urged a visit to the Adirondacks for the purpose of improving mental and physical health (Murray, 1869). The book provided detailed instructions for undertaking such a trip, and thousands of people in urban centers responded. In the same year that *Man and Nature* was published (1864), the *New York Times* ran an editorial about the Adirondacks entitled 'A Central Park for the World' (August 9, 1864). The felling of the forest was of grave concern to those who understood that a healthy economy depended on a healthy ecosystem, and it was also a concern to those who sought out the benefits of nature, simply for respite from an urban way of life.

Throughout the 1870s, some of the most outspoken and influential voices in support of creating a park were those of sportsmen. The famed abundance of fish and game was an enticement that had attracted many from cities into the Adirondack forests. In relation to fishing, the New York State Fisheries Commission had begun stocking programs, but logging, as well as pollution by associated industries, clearly presented a threat (Graham, 1978, pp. 79-80). Adirondack timber was used locally to make charcoal for smelting iron ore, and later to make pulp for paper (Harris and Wilson, 1993). Tanneries were another polluting industry associated with logging (McMartin, 1992, pp. 222-23).

In relation to hunting, white-tailed deer hunting became big business in the Adirondacks. Game laws were few and seldom enforced, and two hunting techniques were especially destructive, 'jacking' and 'hounding'. In jacking, deer were blinded at night with bright lanterns, making them easy targets (Burroughs, 1889). In hounding, dogs would drive deer into lakes where hunters were able to shoot the struggling animals at close distance. Since the Adirondacks were a significant haven for the declining deer herds, there was a great push to provide animals with adequate habitat, even if game protection laws were not at the time effectively enforced. Conservationists' expressed their opinions through the influential magazine, *Forest and Stream*, which first appeared in 1873. Published in New York City, numerous wealthy and powerful men subscribed, and in this way, became aware that the Adirondacks were indeed a significant place, not only for a steady supply of water, but also for the animals that were an essential part of many recreational experiences (Graham, 1978, pp. 82-84).

Contemporary Significance of the Adirondacks

The Adirondacks continued to be a significant region for recreational opportunity throughout the twentieth century. Whereas sportsmen and others seeking relief from the stress of urban living once confined their visits to temporary camps and lavish hotels, an increasing number of vacationers began to build second-homes in the second half of the twentieth century. Some families with vast wealth had established elaborate 'great camps' in the nineteenth and early twentieth centuries (Kaiser, 1982), but these were relatively few. The large number of second-homes in the second half of the twentieth century was far more problematic to the undeveloped character of the Adirondacks.

During this same period, the undeveloped character of the Adirondacks took on increasing significance for its ecological value. Although much of the park remained private, a large area had been dedicated by the state to remain in a natural condition. The idea of a natural area, where ecological processes dominate, became significant regardless of the importance of such areas for maintaining water levels, improving health, or sustaining deer herds. Thanks in part to greater ecological awareness fostered by the writings of Aldo Leopold (1966) and Rachel Carson (1962), the natural world of the Adirondacks was seen as a place for nature itself.

The nearly six million acres that form the Adirondack Park today could easily contain the areas of Yosemite, Yellowstone, Olympic, or Grand Canyon National Parks. It is also larger than several states, including Delaware, Hawaii, Connecticut, and Rhode Island. This vast region provides refuge for 90 per cent of all plant and animal species found in the northeastern United States, including 153 species of nesting birds, 86 species of fish, 54 of mammals, and 35 of reptiles and amphibians. Due to wide variations in elevation, topography, and microclimate, the park includes three biologically distinct regions: temperate deciduous forests, boreal forests, and taiga. About 200,000 acres, or 3.4 per cent of the total park area, has not been cut or burned in the last two centuries (Kudish, 1995). Approximately 2,000 mountain peaks, 1,500 miles of river, 30,000 miles of streams, and 2,800 lakes and ponds are within the park boundaries. Because of its sheer size and diversity, the Adirondacks have been seen envisioned by recent conservation biologists as a 'core' of nature, which might eventually be linked to other cores in Appalachia, northern New England, and parts of Canada by corridors that could facilitate the exchange of wildlife (Medeiros, 1992/93). Second-home development presents a direct threat to the ecological integrity of such a network.

The recreational significance of the Adirondack Park today should not be underestimated. It is within a short automobile ride for 70 million people who engage a wide variety of recreational activities in all seasons. Hiking and camping are very popular, especially in the central mountain ranges known as the High Peaks. This area includes Mt. Marcy, whose elevation of 5,344 feet is the highest in New York. Paddling and boating are undertaken on lakes and rivers, as are hunting and fishing in season. Numerous facilities exist for nordic and down-hill skiing, as well as trails for snowmobiling and ATV-riding. Sightseeing is common at all seasons. Museums, historical sites, golf courses, and theme parks entertain many tourists with diverse interests. In 1999, it was estimated that the Adirondack Park entertained some nine million visitors (Kretser, 2000).

In sum, the significance of the Adirondacks today is in its outstanding natural features that are valued for both biological diversity and recreational opportunity. Biodiversity and recreation can come in conflict, especially as the number of recreationists and/or the technology of the recreational activity increases (Abbey, 1968, pp. 45-67). Particularly when enlarging numbers of recreationists begin to seek out vacation homes, the extent of this tension is exacerbated. In the Adirondacks, such tension is possible because the park is actually a checkerboard of private and public lands. The next section describes how the significance of the Adirondack watershed led to the formation of the New York State Forest Preserve, how both public and private lands were incorporated in the creation of an Adirondack Park, and how the inclusion of both private and public land within the park necessitated a highly unusual degree of state-level land-use planning in order to accommodate ecological values that conflict with recreational demand and second-home development.

The first attempts at land-use control in the Adirondack region date to the 1880s, when the New York State Forest Preserve was established to protect land that had reverted to the state for payment in lieu of taxes. Shortly thereafter, the concept of an Adirondack Park was advanced in order to provide a focus for acquisition of additional property into the Forest Preserve. Although it was simply a line drawn on a map, the patchwork of private and public land contained therein was too heavily tilted in favor of private land for it to ever become wholly state land. Conflicts between public and private were inevitable. Resolution was proposed in the 1960s through the formation of a national park, but various statewide interests opted instead for the establishment of a state agency empowered to control private land within park boundaries. This agency enacts various exercises of the police power to regulate subdivision and development. At the same time, a different agency of state government, in conjunction with nonprofit organizations, continues to expand the Forest Preserve by acquisition of fee and otherwise directs the use of private lands through acquisition of conservation easements.

Formation of the State Forest Preserve and the Adirondack Park

In 1882, destructive and well-publicized flooding of the Mississippi River heightened awareness that flood protection in New York was indeed necessary. In 1883, a severe drought caused water in the principal rivers of the Hudson, Mohawk, and Black Rivers to fall to uncomfortably low levels. In an extensive campaign that beginning in the fall of 1883, the New York *Tribune* argued that the Adirondack forests must be preserved to maintain a steady flow of water. Also in 1883, another source for concern emerged. The Adirondack Railroad Company proposed to build a line directly through the heart of the Adirondack region. As noted by the *Morning Herald* of Utica, if this were to happen, the forest would inevitably disappear (Graham, 1978, pp. 97-99). Other newspapers joined the campaign, and protection of Adirondack watersheds became a major issue almost everywhere south of the Adirondack region (Nash, 1982, pp. 116-121).

Morris Jessup, president of the New York State Chamber of Commerce, argued that it was necessary to save the forests because business within the state would be harmed. He suggested that the state acquire up to four million acres of private land through condemnation by the Governor with the approval of the legislature. This land could be kept as a 'forest preserve', which would insure an adequate supply of water to the Hudson River and the Erie Canal, essential for the shipping of goods. Jessup emphasized that forest destruction had already reduced flow in important waterways, while extensive cutting and fires were certain to follow as the railroad was constructed. State expenditures would be handsomely returned in improved agricultural systems, as well as a better-regulated water supply (Graham, 1978, p. 99).

Jessup's proposal for the state to acquire private land through eminent domain met stiff resistance, especially in the Adirondack region. However, the state already had in its possession approximately 700,000 acres, most of it obtained from private parties who had failed to pay taxes. Even though this land was largely denuded of forest, the emphasis of efforts to protect the Adirondack watershed switched from acquisition of new land by eminent domain to protection of land already owned by the state. On May 15, 1885, Governor David Hill signed into law a bill that established a state Forest Preserve and a three-member Forest Commission. Although 'active' acquisition was explicitly deleted from the legislation, 'passive' acquisition was clearly envisioned, as it was believed that additional properties would revert to the state for failure to pay taxes (Graham, 1978, pp. 100-106). One of the law's more noteworthy provisions is the so-called 'forever wild' clause contained in Section 8:

> The lands now or hereafter constituting the Forest Preserve shall be forever kept as wild forest lands. They shall not be sold, nor shall they be leased or taken by any person or corporation, public or private.

By 1890, a regenerated uproar was directed against railroads and against lumber thieves, who stole timber from the Forest Preserve despite protection afforded by the forever wild clause. Critics viewed the Forest Commission as too submissive. In addition, doctors actively supported additional legislation that would expand what they believed was a great sanatorium. For a decade and a half, invalids from urban areas, especially those suffering from tuberculosis, had been visiting the region in search of a 'wilderness cure' under the tutelage of Edward Trudeau. A physician from New York City, Trudeau believed fresh air and mountain scenery had rid his 'white plague' upon moving to the Adirondack Mountains in 1874. Inspired by Trudeau's example and exhortation, an entire colony of 'cure cottages' arose in the vicinity of Saranac Lake (Cook, 1881; Gallos, 1985).

The idea that the state should actively acquire lands would not go away. Wealthy individuals and clubs were beginning to buy large tracts of land for themselves. For example, the Adirondack League Club, an organization of men predominately from New York City, bought 104,000 acres in the southwestern Adirondacks in 1890. Consequently, for the first time in 1890, the legislature appropriated funds with which the Forest Commission could buy land within counties where Forest Preserve already existed. It gave the Commission $25,000, but stipulated that the highest price to be paid for land was $1.50 per acre, a severe limitation given that the Adirondack League Club paid $4.75 per acre for its purchase. As a collection of separated parcels inexpensively acquired in fee or by tax foreclosure, the Forest Preserve was widely scattered, and the absence of a coherent unity was unsettling at a time when the first national parks were being established in Yellowstone and Yosemite. Earlier that same year (1890), the Commission published a report including a map with a proposed park encircled by a 'Blue Line'. The concept of a park gave the state a defined region for concentrating its land holdings. The idea was to outline a footprint for future acquisition. Since then, the Blue Line has been synonymous with the Adirondack Park (Graham, 1978, pp. 121-122; Terrie, 1997, pp. 97-99).

The effort to create a park was achieved on May 2, 1892, when Governor Roswell Flower signed the Adirondack Park Enabling Act. The law defined the park as 'all lands now owned, or which may hereafter be acquired by the state'. The area within the Blue Line at the time consisted of 2,807,760 acres, of which 551,093 belonged to the state (Graham, 1978, p. 124). Though the Blue Line did not embrace the full extent of the Forest Preserve at that time, it did provide a defined region for future acquisition of private land.

In 1893 and 1894, drought gripped parts of the northeastern United States, and forest fires raged in New York and other states. Business leaders in New York City again became concerned about the health of watersheds. By now, their worries had expanded to include supplies of drinking water and fire protection, as well as waterways for transportation. In the summer of 1894, a convention had been called in Albany to revise the state constitution, and it was here that the New York State Board of Trade and Transportation saw an opportunity to provide the Forest Preserve some clear constitutional protection. The Board of Trade had an active Forestry Committee concerned about watershed issues, and it had also created a special committee to propose constitutional amendments for possible passage at the convention. These two committees collaborated on the following clause, which was passed at the convention by a vote of 112-0 and became part of the Constitution (Article VII, Section 7) when it was approved by the voters of the state later that fall (Donaldson, 1921, pp. 187-96; Graham, 1978, pp. 126-31; Terrie, 1985, pp. 104-108):

> The lands of the State, now owned or hereafter acquired, constituting the Forest Preserve as now fixed by law, shall be forever kept as wild forest lands. They shall not be leased, sold, or exchanged, or be taken by any corporation, public or private, nor shall the timber thereon be sold, removed or destroyed.

The notion of 'forever wild' was clearly carried over from the 1885 legislation creating the Forest Preserve. Only two sentences made up Article VII, Section 7, (renumbered Article XIV, Section 1, at the constitutional convention of 1938), but as part of the New York State Constitution, this clause gave the Forest Preserve protection that cannot be undone by any agency or legislature without consent from all the people of New York in a statewide referendum.

The Adirondack Park, however, contained private land as well as state-owned land. Over the next seven decades, the extent to which the state was able to acquire private land, and hence to consolidate its holdings, was erratic. Little was accomplished, for example, until a bond issue was passed in 1916. Then, in 1919 alone, the Conservation Commission (which had succeeded the Forest Commission) purchased over 97,000 acres mostly in the high peaks, including the summit of Mt. Marcy (Terrie, 1997, p. 128). Despite such efforts, the rate of state acquisition could not keep pace with the possible conversion of private land into second-home developments, particularly once automobiles brought increasing numbers of visitors after the second World War. The demand for second-homes became a significant force of change affecting the Adirondack Park throughout the second half of the twentieth century. The initial response by some park proponents was to recommend the creation of a national park embracing the most scenic features of the Adirondacks.

Creation of The Adirondack Park Agency and the State Land Master Plan

In July of 1967, the New York State Council of Parks released a proposal for a national park within the central portion of the Adirondacks. The proposed Adirondack Mountains National Park would have encompassed 1.12 million acres of state land and an additional 600,000 acres of contiguous private lands, which would have been bought over a fifteen-year period from timber companies and other landowners. The national park would have become the third largest in the United States at the time. It would have included the most noteworthy scenes and geographic features characterizing the Adirondack Mountains, including the High Peaks (Cobb, 2000; Graham, 1978, p. 219).

The suggestion for a national park was met with immediate disapproval by a broad range of groups. The Conservation Department (which had succeeded the Conservation Commission) issued a report contrasting New York State policies with policies at the time for other national parks, such as Yosemite and Yellowstone. Largely on this comparative basis, it was recognized that an Adirondack Mountains National Park would both disorganize the timber industry and dislocate much of the tourist industry. The timber industry also objected because the proposal included some of the densest private forest lands in the state. Hunters balked because they were not usually allowed in other national parks. In the view of conservationists, a ban on hunting could also have unforeseen ecological effects (Cobb, 2000; Graham, 1978, p. 222).

The Association for the Protection of the Adirondacks, as well as other conservation organizations, lobbied relentlessly. One of their main arguments was that federal control could possibly compromise the 'forever wild' guarantee afforded by the state constitution (Cobb, 2000). Administrators of national parks had never hesitated to cut down trees, open roads, and erect large hotels for tourists, if it meant visitors would have easier access to the regional attractions (Graham, 1978, p. 224). Under federal ownership, the Adirondack region would not have such indisputable protection. The national park proposal slipped from the realm of possibility late by 1967. Though unsuccessful, the proposal for an Adirondack Mountains National Park paved the way for more stringent land-use planning by the state.

With second homes booming in adjacent Vermont, and local land-use planning within the Adirondack Park rudimentary at best, Governor Nelson Rockefeller appointed in 1968 a Temporary Study Commission on the Future of the Adirondacks. Harold Hochschild, whose estate would have been subsumed by the proposed national park, was appointed chairman. Among other things, the commission found that land speculation and unplanned development on private land were imminent and potentially devastating to the open space of the park. Moreover, state agencies, particularly the newly organized Department of Environmental Conservation (DEC), had powers that were restricted to management of public land only. The first of 181 recommendations in the Commission's report, *The Future of the Adirondack Park* stated that, '[a]n independent, bipartisan Adirondack Park Agency should be created by statute with general power over the use of private and public land in the Park' (Temporary Study Commission on the Future of the Adirondack Park, 1970, p. 9). Among other notable recommendations was that Article XIV, Section 1 of the state constitution should remain unchanged, and that 'there should be no participation by the federal government in the management of state or private land in the Adirondack Park' (Temporary Study Commission on the Future of the Adirondack Park, 1970, p. 24).

With its first recommendation, the commission was advising the state to assume a portion of the responsibility for regulating private land within the park. The commission believed that local governments, with large areas and small populations, did not have sufficient resources to create effective comprehensive master plans. Although the commission advocated a partnership between local communities and the proposed Adirondack Park Agency (APA), a definite shift in the locus of land-use planning was clearly envisioned. The commission recommended that APA prepare a comprehensive plan for the entire park, and that it implement regulatory power through the establishment of a permit system. As for state lands, APA would have planning authority consistent with 'forever wild' and subject to consultation with DEC, which would continue to administrate the Forest Preserve (Temporary Study Commission on the Future of the Adirondack Park, 1970, pp. 30-31).

On May 10, 1971, Governor Rockefeller sent a bill to the legislature that would create the APA. By June 7, the Assembly had passed it, 123-24, and on the next day, the Senate also approved it, by a vote of 23-14 (Graham, 1978, pp. 242, 246). As enacted, the APA board was to be composed of eleven commissioners, three of which were required to be the Commissioner of DEC, the New York Commissioner of Commerce, and the New York Secretary of State. The other eight were to be private citizens, of which five were to be park residents. Citizens were to be appointed by the governor for staggered terms, with no more than five from one political party. Although this board would have final say on matters of broad policy, most actual implementation of park planning would be accomplished by a staff of approximately 70 persons, including specialists in national resources, land-use planners, environmental designers, lawyers, and other professionals. The first task of the new agency was to develop two master plans for the park, one for the Forest Preserve and the other for private land.

By June of 1972, working at a frantic pace, APA completed the *State Land Master Plan*, developed in consultation with DEC. This plan placed all state-owned lands and rivers into several categories, 'according to their characteristics and capacity to withstand use', and it also provided 'general guidelines and criteria for the management and use of lands within such classifications' (Adirondack Park Agency, 1972, p. 1). Although logging on state lands was prohibited by creation of the Forest Preserve, as well as by passage of the constitutional amendment establishing 'forever wild', recreation remained a major use. Recreational demand soared throughout the twentieth century, in terms of both the number of recreationists and the variety of recreational activity. The significance of the Adirondack Park as a resource for recreation cannot be overstated, and both APA and DEC were fully cognizant of it when developing a classification scheme for land and water owned by the state. The resulting categories were:

- *wilderness:* 'an area where the earth and its community of life are untrammeled by man – where himself is a visitor who does not remain...[it] is further defined to mean an area of state land or water having a primeval character, without significant improvements or permanent human habitation'

- *primitive:* '[e]ssentially wilderness in character but (a) which contains structures, improvements, or uses that are inconsistent with wilderness...and/or (b) which contains, or is contiguous to, private lands that are of a size and influence to prevent wilderness designation'

- *canoe:* 'where the watercourses or the number and proximity of lakes and ponds make possible a remote and unconfined type of water orientated recreation in an essentially wilderness setting'

- *wild forest:* 'where the resources permit a somewhat higher degree of human use than in wilderness, primitive or canoe areas, while retaining an essentially wild character'

- *intensive use:* 'provides facilities for intensive forms of outdoor recreation by the public,' including campgrounds of more than 20 sites and developed beaches

- *wild river:* 'free of diversions and impoundments, inaccessible to the general public except by water, foot or horse trail and with a river area primitive in nature and free of man-made development'

- *scenic river:* 'free of diversions or impoundments except for log dams, with limited road access and with a river area largely primitive and undeveloped or which is partially or predominately used for agriculture, forest management and other dispersed human activities'

- *recreational river:* 'readily accessible by road or railroad, that may have development in the river area and that may have undergone some division or impoundment in the past'

- *travel corridor:* 'that strip of land constituting the roadbed and right-of-way for state and interstate highways in the Adirondack Park and those lands immediately adjacent to and visible from these highways' (Adirondack Park Agency, 1972, pp. 8-18).

After nine public meetings, attended by approximately 1000 people, the *State Land Master Plan* was submitted to Governor Rockefeller, who subsequently approved it, thereby enacting it as state policy. Almost one million acres of Forest Preserve were placed into 15 areas of wilderness. Non-conforming uses, such as roads, powerlines, and trails for motorized vehicles, had to be removed from these areas by the end of 1975. Motorized vehicles were permitted on more than one million acres comprising 15 areas of wild forest. Intensive use included two ski facilities operated by the state, as well as major boat-launching sites and campgrounds. One canoe area was established, containing over 50 water bodies embedded within 18,000 acres. Because it offered something for everyone, the *State Land Master Plan* was not very controversial. After the nine public hearings, only slight changes were made in the initial map and text of the plan (Graham, 1978, pp. 247-48). See Figure 8.1.

Creation of the Private Land Use and Development Plan

The next task of the APA was to develop a master plan for use of private lands. In accordance with principles articulated by Ian McHarg (1969), the process was based on natural resource considerations that dictate which areas are more or less 'intrinsically suitable' for development. Inventories were made of watersheds, soil, wetlands, slope (topography), scenic resources, and significant biological areas, such as habitat for unusual species. These inventories were compared with

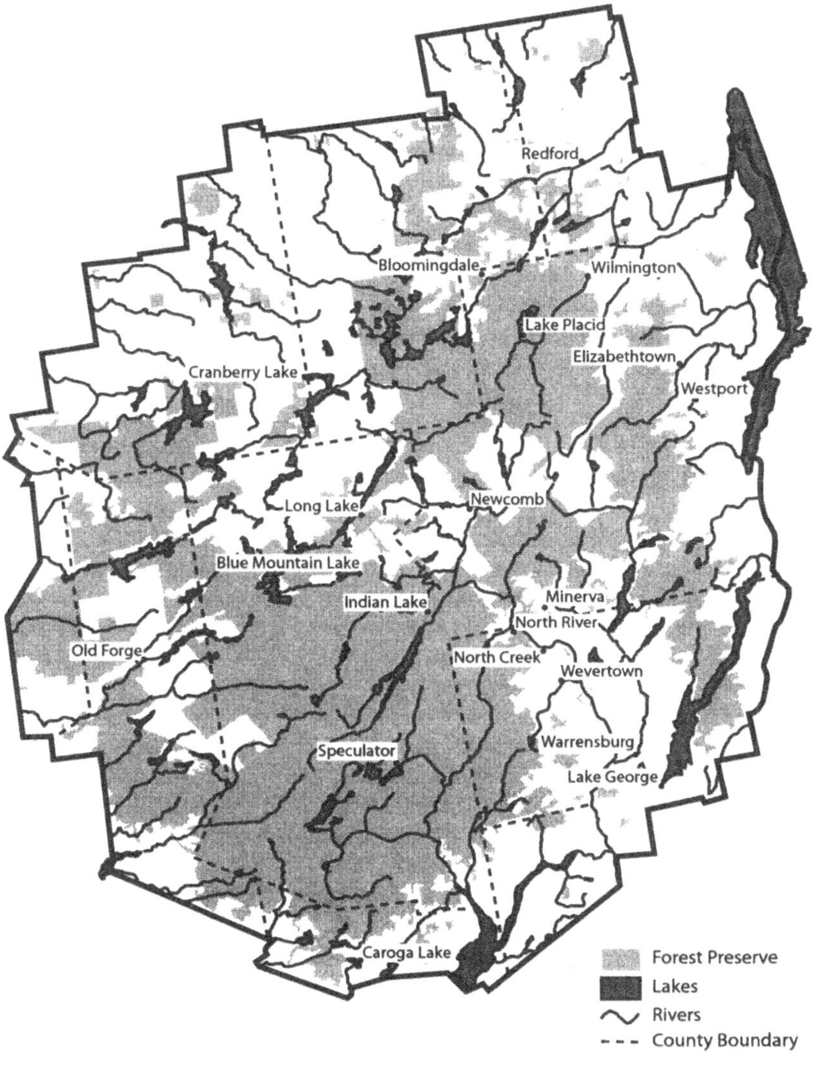

Figure 8.1 Adirondack Park Plan

existing land-use, so that future development would not only be intrinsically suitable, but compatible with existing patterns of development on private land.

This resulting plan was composed of a map and text that classified private lands into six categories according to both type and level of development that would be permitted in light of natural resource limitations. The six categories were:

- *hamlet:* 'range from large, varied communities that contain a sizable permanent, seasonal and transient population....to smaller, less varied communities...Hamlet areas will serve as the service and growth centers of the park'

- *moderate intensity use:* 'areas where the capability of the natural resources and the anticipated need for future development...primarily residential in character, is possible, desirable, and suitable'

- *low intensity use:* 'readily accessible areas, normally within reasonable proximity to a hamlet, where the physical and biological resources are fairly tolerant and can withstand development at an intensity somewhat lower than found in hamlets and moderate intensity use areas'

- *rural use:* 'areas where natural resource limitations and public considerations necessitate fairly stringent development constraints. These areas are characterized by substantial acreages of one or more of the following: fairly shallow soils, relatively severe slopes, significant ecotones, critical wildlife habitats, proximity to scenic vistas or key public lands'

- *resource management:* 'lands where there is need to protect, manage and enhance forest'

- *industrial use:* 'areas (1) where existing land uses are predominantly of an industrial or mineral extractions nature or (2) identified by local state officials as having potential for new industrial development' (Adirondack Park Agency, 1973, pp. 8-12).

The *Land Use and Development Plan* (LUDP) is essentially a large-scale zoning and subdivision ordinance. In addition to listing conforming uses within each category, the LUDP also established 'guidelines for the overall intensity of development' within four of the six categories. Intensity guidelines were limits on lot size expressed as the number of principal buildings per square mile. When translated into average lot size, 500 buildings per square mile in moderate intensity became 1.3 acres per building, 250 in low intensity became 3.2 acres per building, 75 in rural use became 8.5 acres per building, and 15 in resources management became 42.7 acres per building. No intensity guidelines were established for hamlet and industrial use. The APA gave itself little jurisdiction over hamlets in the LUDP. All uses of land were 'considered compatible with the character, purposes and objectives of hamlet areas' (*Land Use and Development Plan*, 1973, p. 8). In this way, the LUDP did not supersede land-use planning by local governments in villages throughout the park.

Beyond the above provisions, additional restrictions were placed on proposed development in the vicinity of shorelines. Minimum widths were established for lots, to restrict the number of buildings that would eventually ring a water body; minimum set-backs were provided between building and shore, to protect water from seepage of septic effluent; minimum shoreline frontages were required when non-waterfront lots were given deeded or contractual access to the water body; and restrictions were placed on removal of vegetation, to ensure the natural character of shorelines. Specific standards varied as a function of land-use category, with more stringent requirements in the more restrictive categories of resource management and rural use (Adirondack Park Agency, 1973, p. 15).

The permitting process under the LUDP is fairly typical. Once a completed application is received, APA has 90 days to approve the project and issue a permit, with or without conditions, or 60 days to notify the sponsor that a public hearing will be held. Public hearings are held on only about five per cent of all applications submitted to the Agency. Most projects are approved by APA staff, but those involving a subdivision of more than 50 units, those requiring a variance, and those for which a public hearing has been held require approval by the eleven members of the APA commissioners (Adirondack Park Agency, n.d., pp. 16-19). As depicted in table 8.1, very few applications have been disapproved or eventually withdrawn in recent years.

Finally, the LUDP divided proposed development into two classes: A and B. Class A was made up of large projects having a potential regional impact on a wide area of the park. Class B projects were primarily of local effect. Interestingly, a class B project in one land-use category is a class A project in a more restrictive category. For example, a tourist accommodation is considered a class B project in moderate intensity and low intensity, but it is a class A project in rural use. Similarly, commercial sand and gravel extraction is classified as a class B project in moderate intensity, low intensity, and rural use, but it is class A if located in resource management (Adirondack Park Agency Authority, n.d.). In this respect, the sliding scale represents an inconsistency of sorts. After all, to some observers, either a project has regional significance or it does not. Nonetheless, the purpose of the two classes was to distinguish between uses that might eventually be determined by local governments and uses whose determination would be retained by APA under all circumstances. The LUDP allowed that class B projects could be regulated by individual local governments once a land-use plan had been prepared and approved by APA. In other words, this provision was designed as an incentive to promote the implementation of comprehensive master plans by villages and towns within the park.

Despite this provision, year-round residents of the park were not attracted to the LUDP. In fact, their reaction, once Governor Rockefeller signed the LUDP into law in May of 1973, was one of outrage. Given that less than 10 per cent of the land was zoned by local government at the time the LUDP was enacted (The

Temporary Study Commission on the Future of the Adirondack Park, 1970, p. 27), most permanent residents felt inalienable rights had been usurped. From the local perspective, private property was thought to be something one could do with as one wished. It was bad enough that the state owned so much land as Forest Preserve; now, the heavy hand of New York State was coming down, as well, on private land within the park. The proportion of private land allocated to each category in the LUDP gave the appearance of making things worse for the average citizen than they actually were. Most of the land (about 80 per cent) was placed in the two most restrictive categories: about 1.5 million acres in resource management and one million acres in rural use. However, much of this land was owned by large paper companies, with little interest in development at the time.

Table 8.1 Recent permit activity by the Adirondack Park Agency

	Total Number (1990-1999)
Applications Received	3976
Applications Ruled Non-jurisdictional	309
Permits Issued	3242
Applications Disapproved	5
Applications Withdrawn	177
Applications Pending	243

Source: Bauer, Peter, *Growth in the Adirondack Park: Analysis of Rates and Patterns of Development* (North Creek, NY: The Residents' Committee to Protect the Adirondacks, 2001), p. 27.

Nonetheless, the locals did have a point. Zoning regulations that require lots of 42.7 acres per principal building (resource management) and 8.5 acres per principal building (rural use) were highly unusual thirty years ago. For the strictness of its regulations, as well as the amount of land placed under regulation, the LUDP was an extraordinary exercise of land-use control. In response, permanent residents organized themselves into citizen groups, the most vocal of which was the League of Adirondack Citizens' Rights. They issued newsletters, such as the *Adirondack Defender,* and they held numerous meetings with an unwavering intention to 'abolish the APA' (Graham, 1978, pp. 254-57; Terrie, 1997, pp. 168-69). Despite the predominate feeling of permanent residents that the regulations were confiscatory, the LUDP has been upheld as a proper exercise of the police power that neither imposed an unjustified 'taking' nor violated 'due process'.

How the Adirondack Park is Defined

The Adirondack Park can be defined in terms of the physical attributes of its landscape. It can also be defined on the basis of social characteristics and cultural circumstances. When most people are asked about the Adirondacks, they tend to think about the region without its social and cultural component. Throughout the nineteenth century, the local Adirondack economy had a significant agricultural dimension; however, visitors who created imagery of the Adirondack landscape obscured this reality. In other words, people outside the Adirondack region established a different definition of the Adirondack Park than people who lived within it. Different definitions of the same place would eventually exacerbate the difficulties of managing a park that contained both public and private land.

The Adirondack region is often defined merely in the language of physical geography. It can be described, for example, as the region north of the Mohawk River, south of the St. Lawrence River, west of Lake Champlain, and east of the TugHill Plateau. In recent articles describing biodiversity, the park was defined as 'an oasis of wilderness' with 2,800 lakes, ponds, 42 peaks over 4,000 feet, 2,000 miles of hiking, 1,500 miles of wild rivers, 30,000 miles of brooks and streams, and 43 state campgrounds. (Medeiros, 1992/93; Pasquarello, et al. 1994). Yet, 'Adirondacks' summons up a variety of images: busy centers of tourism, such as Lake Placid; remote wilderness areas, with remnants old-growth forest; the poverty of a deteriorating village, where a paper mill has recently gone out of business; a cottage nestled in the pines, with a canoe or boat tethered to a dock; dilapidated barns and abandoned farmhouses. The Adirondack chair has become a symbol of leisure recognized across the United States. Historian Philip Terrie has noted that '[t]he Adirondack region today is a complex and diverse place, existing both in the geographical reality of the physical landscape and in the imaginations of the millions of people who have lived in, worked in, traveled through, or thought about it...People tell stories about the land that reflect their needs. They project their needs onto the land in the stories they tell about it. They define – in a sense, create – the land in their stories...When a story becomes widely accepted, it helps to promote certain attitudes toward the land and thus eventually contributes to the formation of public policy' (Terrie, 1997, pp. xvii-xviii).

Few people think of fertile fields or rich agricultural tracts when they think 'Adirondacks'. However, the first visitors in the mid-nineteenth century described both the progress and possibilities of farming in the region. Early publications of the Adirondack region foresaw a future based largely on agriculture (e.g. Todd, 1845; Watson, 1853). Even painters of the Hudson River School, who entered the Adirondack interior in order to find inspiration and create landscape art, inadvertently included agriculture as part of their vision. Clearings in the Adirondack forest became an attractive perspective from which to develop romantic ideas of wilderness. (McMurry, 1999). While there were clear limits on agricultural cultivation in the Adirondacks, farming nonetheless remained an important element in the region's economy throughout the nineteenth century. In fact, most people who lived in what would become the Adirondack Park were very likely farmers (Harris, 1999).

Nearly all Adirondack families made agriculture their primary occupation, though many had to combine farming with various other jobs. Agriculture at the time was largely for subsistence, and it soon became an unavoidable aspect of Adirondack life that no one could survive on farming alone. Residents needed income to purchase essentials like stoves, certain items of clothing, and equipment for hunting, fishing, and farming. One common way to earn income was through trapping and the sale of furs; another was to manufacture items in the home for sale to others, such as butter, split shingles, or knitted clothing (Terrie, 1997, pp. 32-33). Many farmers were also loggers in winter. A considerable amount of money could be garnered through seasonal employment in the woods (Harris, *et al.*, 2000).

Eventually, other trades were developed that increasingly defined life and work for people who resided in the Adirondack Park. Greater amounts of time for leisure among people who lived in urban areas brought writers, artists, sportsmen, and other recreationists into the area. This, in turn, generated a need for guides and hotels, with attendant services. The arrival of tourists was a significant development for Adirondack people. A tourist-based economy would eventually become essential to their livelihoods, and it would also require a change in the way they related to the Adirondack landscape. The arrival of tourists meant the end of an era marked by farming and relative self-sufficiency. As tourists made their imprint, the Adirondacks that local residents had known and the manner in which they defined their surroundings began to disappear (Terrie, 1997, pp. 40-41).

Agriculture reached its peak in the region around 1890, but by this time, ideas of what constituted the Adirondack region were already being transformed into terms embodied in the history and legislation of the Forest Preserve and the Adirondack Park. As noted in a recent study of farming in the Adirondack Park, '[a]griculture became obscure in representations of the Adirondacks because of shifts in social, cultural, and economic conditions which permitted the reconstruction of the Adirondack imagery to fit the views of an urbanizing culture and its leisured elite' (McMurry, 1999 p. 119). As described earlier, the creation of the Forest Preserve and the Adirondack Park resulted from a flurry of policy-making by leaders of government and commerce in the 1880s and 1890s. The key authors of this legislation and the voices that prevailed were almost exclusively from outside the region. To lumber-barons, the Adirondacks were timber for exploitation; to business leaders downstate, the Adirondacks were critical watershed in need of protection; to the infirm, the Adirondacks were a sanatorium for mental and physical health; and, increasingly, to the growing number of hunters and other vacationers, the Adirondacks were a haven for outdoor recreation. Each interest wanted to protect its idea of what constituted the 'Adirondacks' and each, in turn, helped to define the Adirondack Park as it is known today (Terrie, 1997, p. 83). What the Adirondacks meant to year-round citizens was largely lost. It became subservient and subordinate to others.

Through stories heard from their guides as well as accounts of recreational experiences, romantic writers and artists constructed an image of the Adirondacks that has largely determined popular attitudes to this day. The Adirondack landscape became 'wilderness'. Ignoring the villages where they stayed and overlooking the lives of residents who fed and housed them, tourists throughout the twentieth century reinforced a story about the Adirondacks that defined the area without a significant human presence. Although this definition suited the desire for a natural and picturesque landscape, it neglected the perspective of permanent residents (Terrie, 1997, pp. 54-55). In the same way that framers of the Adirondack Park failed in 1892 to grasp fully the implications of enclosing so much private land within the physical boundary of the Blue Line, urbanities who gave the Adirondack Park its cultural definition did not fully understand that they were omitting the way local people defined the region as home and source of livelihood. As illustrated in the controversy over the LUDP, conflicts have endured ever since.

Forces of Change Affecting the Adirondack Park

During the past thirty years, many natural and cultural processes have affected the Adirondack Park. This section reviews only a handful, focusing especially on those that have produced current controversies in park planning and management. The first force of change has been the steep rise in recreational use and second-home development. In this context, it is also instructive to examine the impact, or lack thereof, of such an increase on socio-economic conditions for park residents.

Forces of change affecting natural attributes of the Adirondack Park include acid rain and invasive or exotic species. These processes, like recreational demand and second-home development, originate outside the park. Increases in affluence, leisure time, mobility, and transboundary pollution are deeply rooted in global events far beyond the scope of this chapter. This section sacrifices consideration of causes of change to focus on change itself. With limited ability to influence underlying causes, park managers must, with few exceptions, limit themselves to dealing with the manifestations of these changes.

Second-Home Development

After World War II, the steep rise of autotourism led individual entrepreneurs, corporations, and local chambers of commerce to compete with one another to create tourist attractions, such as theme parks, designed to lure tourists and their dollars. The prospect of autotourism led to construction of the Northway, a stretch of Interstate 87 between Albany and the Canadian border. To pave the way, New York voters were asked in 1959 for approval of an amendment to the New York State Constitution that would permit the highway to run through approximately 300 acres of the Forest Preserve in the eastern portion of the Adirondack Park. Business and residents of central Adirondack towns believed future prosperity depended on easy access for tourists. In this instance, their perspective prevailed, allowing New York to secure federal funds established by the Interstate Highway Act of 1956. The completion of the Northway in 1967 made traveling into the region much easier and faster than ever before (Graham, 1978 p. 212; Terrie, 1997, p. 162).

Completion of the new highway had many ramifications: it allowed an unprecedented number of vacationers to visit the area, it increased recreational demand, and it resulted in more visitors deciding that the Adirondacks would be a pleasant place to have a summer cottage. With millions of acres of land under private ownership, and miles of lake and river shores available for purchase, several important developers intended to build major second-home projects in the park. The Great Northern Capital Corporation of Toronto, for example, bought 10,000 acres in Herkimer County for the purposes of damming the Moose River, constructing golf courses and marinas, and selling the property for 5,000 second homes (Terrie, 1997, p. 167). In 1972, the Horizon Corporation of Tucson purchased 24,000 acres in St. Lawrence County. This company envisioned as many as 10,000 houses, access roads, a golf course, skiing facilities, and several dams on the Grasse River. A third proposal was the Ton-Da-Lay project, involving 4,000 units for approximately 20,000 people on 18,500 acres of forested land in Franklin County (Graham, 1978, pp. 248-49). Conservationists objected, of course, to such large subdivisions, stressing the importance of having a strong regional land-use plan to control developments of that size. Ultimately, the LUDP was passed by the New York State Legislature in 1973, just before Horizon began actual construction, and none of these projects ever came to fruition.

After the economic slow-down of the late 1970s, second-home development again picked up during the 1980s. Sales of subdivided property tripled between 1982 and 1985, and by 1988 had doubled again (The Commission on the Adirondacks in the Twenty First Century, 1990, pp. 47-48). As a consequence, Governor Mario Cuomo established another temporary study commission in 1989 to study the status of the Adirondack Park and the effectiveness of APA. The report of this temporary commission focused primarily on strategies for arresting the decline in open space.

Against a backdrop of rising demand for second homes, it is informative to examine demographic statistics for permanent residents. Two of the 62 counties in New York are located entirely within the Blue Line: Essex and Hamilton. These counties currently account for about 34 per cent of 130,000 permanent residents presently in the park. They also contain some of the principal tourist attractions, including the High Peaks, Mt. Marcy, the Adirondack Museum, and the village of Lake Placid. Yet, despite the appeal of these counties to visitors from outside the park, the number of permanent residents has remained relatively static throughout the past century (see table 8.2).

Table 8.2 Resident population of Essex and Hamilton Counties in the Adirondack Park

County	Thousands of People				
	(1910)	(1930)	(1950)	(1970)	(1990)
Essex	33.5	34.0	35.1	34.6	37.2
Hamilton	4.4	3.9	4.1	4.7	5.3

Source: U.S. Bureau of the Census, *1990 Census of Population and Housing*; material compiled by Empire State Development, State Data Center.

According to the most recent statistics available, the population of Essex County on July 1, 1999 was 37,507 people, an increase of less than one per cent since 1990, and Hamilton County was 5,190, a decline of almost two per cent (Nelson A. Rockefeller Institute of Government, 2000, p. 7). The shocking absence of change in the number of permanent residents is significant. Population on private lands within the park has stagnated, and despite the remarkable growth in recreational use and second-home development, so have economic conditions and attendant public services, like education and health care. For 1999, unemployment was 8.6 per cent in Essex County and 10.6 per cent in Hamilton County, compared to 4.2 per cent for the rest of the state (Nelson A. Rockefeller Institute of Government, 2000, p. 90). According to the final report of the Cuomo Commission, per capita income for the entire park in 1985 was $8,459, compared to $9,029 for rural counties elsewhere in New York and $11,765 statewide (The Commission on the Adirondack Park in the Twenty First Century, 1990, p. 29).

Primary and secondary schools in the Adirondack Park range in size from only six children to a few hundred. Per pupil spending for public school districts in 1987 was $5,450, compared to a statewide average of $6,564 outside New York City (The Commission on the Adirondack Park in the Twenty First Century, 1990, p. 37). The community college system is small and under-funded. Until 1997, no four-year college was located within the Adirondack Park, and in 1990, only 8.6 per cent of adult residents in the park had earned a bachelors degree. Many young people who become educated move out of the area. The percentage of graduating high school students who leave their home communities surpasses the state average overall. The reason for such outmigration is lack of local employment opportunities. Most Adirondack towns and villages have few jobs for high school or college graduates (Brooks, 1999).

The Cuomo Commission noted that local communities with a large proportion of seasonal homes were penalized in several ways. First, these communities received less state funding in relation to the amount of developed land, since standard state-aid formulas are based largely on permanent population, excluding seasonal residents (The Commission on the Adirondack Park in the Twenty First

Century, 1990, p. 29). Second, strong demand for second homes drove up the price of land and housing in some communities. Specifically, the report stated that '[i]n establishing and promoting the Park, the state has contributed to a strong market for recreational homes....this, in turn, has tended to increase land and housing costs for residents, a trend that will only grow stronger in the future' (The Commission on the Adirondack Park in the Twenty First Century, 1990, p. 32). Special measures would be required to make housing affordable for park residents.

The commission proposed the creation of an Adirondack Park Community Development Corporation, which would have limited bonding authority to provide affordable housing and infrastructure in hamlet and moderate intensity areas. It also recommended that the state should provide technical support through the proposed Community Development Corporation to assist local governments in obtaining state and federal funds for public housing and other development projects. In addition, the commission suggested that builders be required to include a certain percentage of affordable housing in subdivisions for residential development. (The Commission on the Adirondack Park in the Twenty First Century, 1990, p. 33).

Despite such recommendations to reduce the cost of housing, as well as other suggestions to improve education and to promote the economy through expansion of forestry and tourism, the majority of park residents were convinced that the Cuomo Commission did not represent their interests. They responded with disfavor and protest, probably because bitterness still lingered after implementation of the LUDP. On this occasion, in contrast to events 20 years earlier, political resistance by permanent residents overshadowed more stringent protection of the Adirondack Park. As a result, none of the 245 recommendations offered by the commission had been enacted by 1994 (Pasquarello, Buerger, & Randorf, 1994), and only a few dozen had been adopted by 2001 (Bauer, 2001, p. 16).

Acidic Deposition

Acidic deposition has emerged as a critical issue in the Adirondack Park during the past 25 years. Gaseous emissions of sulfur dioxide and nitrogen oxides react with moisture to form sulfuric and nitric acids in rainfall, snow, fog, mist, and cloud moisture. Approximately two-thirds of sulfur dioxide arises from electrical generating stations, predominantly coal-fired power plants. Highest emissions originate in mid-western states of the Ohio Valley. The two major sources of nitrogen oxides are power plants, using any fossil fuel, and gasoline-powered vehicles, particularly cars and trucks. Prevailing winds carry gaseous emissions from the midwest toward the northeast, and conversion to acidic substances occurs en-route. Rainfall in summertime tends to be more acidic than precipitation in winter (McCormick, 1997). While unpolluted rainfall is slightly acidic, due to the presence of carbon dioxide in the air, wet deposition in the Adirondack region is among the most acidic in the United States, with pH values averaging 4.5 to 4.6, 150 times more acidic than distilled water (National Atmospheric Deposition Program, 2000).

Mountainous areas of the Adirondack Park are specifically vulnerable to acidic deposition for several reasons. Higher elevations intercept air masses, especially in the southwestern portion of the park, and produce abundant precipitation. Acidity in cloud moisture enveloping mountains is often greater than acidity of rainfall at lower elevations. Soils in the Adirondacks are thin and have low capacity to neutralize acidic inputs. Furthermore, watersheds at high elevation are often relatively small in size, so less soil is present to neutralize acidic deposition (Park, 1987).

Acidic deposition has subtle, incremental, and long-term effects that threaten the ecological health of the park. Scientific studies in the 1980s, under the National Acidic Precipitation Assessment Program funded during the Reagan administration, focused mostly on impacts to lakes. Aquatic organisms have varying abilities to tolerate acidity. Rainbow trout, small-mouth bass, crayfish, mussels, salamanders, and some plankton are among species least able to adapt to increasing acidity. Early life-stages of many organisms are more sensitive to high acidity than adults. For example, fish and amphibian eggs, fish fry, and tadpoles cannot survive even brief periods of high acidity. Many species decline in number and eventually disappear (U.S. Environmental Protection Agency, 1999; Roy, *et al.*, 2000).

In the mid 1980s, two different surveys found that between 12 and 24 per cent of lakes in the Adirondack Park were 'seriously' acidic (Roy, *et al.*, 2000). Hundreds of lakes are chronically acid, and from 50 to 70 per cent are potentially susceptible to episodic acidification, that is, a rapid increase in acidity associated with melting of the snowpack in springtime (U.S. Environmental Protection Agency, 1999). By 2000, a total of 41 per cent of lakes in the Adirondack Park were either chronically or episodically acid (Driscoll, *et al.*, 2001).

In addition to affecting surface waters, acidic deposition has altered chemical composition of soils, affecting the balance of major plant nutrients. Hydrogen, sulfate, and nitrate in acid precipitation have contributed to the leaching of calcium and magnesium to depths below the reach of plant roots. Acidity also causes potentially toxic metals, such as aluminum, to become more soluble in soils and to drain into lake waters. Aluminum in soil water harms small roots and interferes with their ability to absorb calcium (Krajick, 2001). Aluminum is also toxic to fish. While aluminum is mobilized from local soils, mercury often accompanies sulfur and nitrogen emissions because trace amounts are found in coal burned in power plants. Certain forms of organic mercury can bio-accumulate in fish. DEC has issued mercury consumption advisories for selected lakes in the Adirondacks, so that anglers should not eat fish they catch.

Long-term effects of acidic deposition on trees in the Adirondack Park has been another area of concern. More than half of larger red spruce at high elevations, for example at Whiteface Mountain, died during the 1970s and 1980s (Craig and Friedland, 1991). Some mortality of red spruce has also been observed at lower elevations in the western Adirondacks (Shortle, *et al.*, 1997). Acidic deposition leaches a small, but crucial portion of calcium from needles of red spruce. Removal of calcium leaves the youngest needles less able to withstand extreme cold, so these needles are more susceptible to death by freezing in winter (DeHayes, *et al.*, 1999). Decline of sugar maple has also been observed in the northeast since the 1960s, but dieback is thought to be from several causes, including insect infestations and droughts. Acidic deposition may contribute to these stressors by depleting calcium and magnesium from soils, especially for trees living on soils already poor in nutrients (Driscoll, *et al.*, 2001).

New York State has attempted to reduce sulfur dioxide emissions from sources within its borders. Although it has only about a dozen, small, coal-fired power plants, the state passed its own acid deposition control act in 1984, the first such legislation in the country. This action, while useful for states downwind of New York, was largely symbolic for addressing acidic deposition in the Adirondack Park. According to a DEC study, approximately 83 per cent of the sulfur deposition occurring in the southwestern Adirondacks comes from outside the state (Roy, *et al.*, 2000).

No federal legislation dealt with acidic deposition from power plants until the United States Congress amended the Clean Air Act in 1990 to include the Acid Rain Control Program. This program set a national ceiling on sulfur dioxide emissions, required a reduction of nitrogen oxide emissions, and created a market-based system that would allow utilities to trade allowances for emissions of sulfur dioxide. When the program is fully implemented in 2010, national emissions of sulfur dioxide should be reduced by about 10 million tons (40 per cent), compared to 1980 levels, and nitrogen dioxide emissions should be reduced by 2 million tons per year. To date, power plants have curtailed sulfur dioxide emissions faster than required, and at much less cost than originally estimated, by either industry or the federal government. Concentrations of sulfur dioxide in the air, and amounts of sulfate in deposition, declined from 36 to 47 per cent between 1988 and 1997 (U.S. Environmental Protection Agency, 1999). Nonetheless, a recent report by ten prominent researchers concluded that utilities would need to cut sulfur dioxide by an additional 80 per cent, from the levels set by the 1990 amendments, if waters and soils in the Adirondack region are to recover from the cumulative effects of decades of acidic deposition (Driscoll, *et al.*, 2001).

Efforts by New York State to protect the Adirondack Park from acidic deposition continue. In the fall of 1999, Governor George Pataki directed electrical power generators in the state to curtail sulfur dioxide emissions by an additional 50 per cent beyond reductions currently mandated by the federal government (Roy, *et al.*, 2000). Members of New York State's Congressional delegation continuously attempt to legislate improvements at the federal level. In March of 2001, for example, Representative Sherwood Boehlert (R-NY) introduced a bill calling for both a 75 per cent decline in sulfur dioxide below present requirements of the Acid Rain Control Program, and a 75 per cent decrease in nitrogen oxide from recent emission levels (Krajick, 2001).

Invasive or Exotic Species

Like acid deposition, global warming through the build-up of atmospheric greenhouse gases is a force of change affecting plants and animals in the Adirondack Park. And, like acid deposition, the source of this problem is both anthropogenic and well outside the boundaries of the park itself. Global warming will gradually result in climatic conditions favorable to a different set of plants and animals than those which have occupied the Adirondack landscape in the recent past. As a force of change, global warming will be even more incremental and less dramatic than acid precipitation. Species unable to adapt will disappear, and species that can adapt will flourish. New flora and fauna will invade the Adirondacks, similar to bait inadvertently released by anglers, aquatic vegetation introduced by transient watercraft, and plants that escape from domestic gardens. Are these exotic species that need management?

The Adirondack Nature Conservancy (ANC) is a private, non-profit organization interested in the integrity of natural communities in the Adirondack Park. Established in 1971, its 'mission is to preserve plants, animals, and natural communities that represent the diversity of life on earth by protecting the lands and waters they need to survive'. Like other Nature Conservancies throughout the world, ANC has emphasized preservation of critical biological habitat. In 1993, ANC acquired easements on 1,900 acres of land containing abandoned mines that had become an area of hibernation for an estimated 119,000 bats. This area is the largest 'bat hibernaculum' in the northeastern United States, accommodating all six species of bats that reside in the Adirondack Park. ANC has also purchased, or received as gifts, parcels of land containing unusual plant communities, including bogs and pine barrens. For land under its ownership, ANC operates an active stewardship program to maintain and perpetuate desired species.

Given its orientation, it is not surprising that ANC should be worried about threats to plants and animals. Whether an indirect result of chemical and hydrological changes brought about by acid deposition and global warming, or a direct result of introduction by careless anglers, boaters and gardeners, invasive species have the potential to create serious perturbations in natural systems. In recent years, exotic species have become a growing emphasis of ANC (Spada and Brown, 2000; Schoch 2001).

An example of an invasive aquatic species is Eurasian watermilfoil. It is usually introduced to a water body from the bottoms of watercraft, which have collected it elsewhere. Under favorable conditions of light, temperature, and nutrient availability, Eurasian watermilfoil increases rapidly. By blocking light from low-lying vegetation, dense layers of this non-native species inhibit the ability of certain native aquatic plants to survive. Once established, containment has proven quite difficult and expensive. A variety of methods have been employed, such as mechanical harvesting, herbicides, and biological controls. However, the best ways to manage any exotic species is to prevent its introduction in the first place and to increase public awareness regarding the importance of prevention (New York State Department of Environmental Conservation, 1997).

Beyond Eurasian watermilfoil, ANC recently completed a survey of terrestrial invasives, especially along roadways. The survey produced a database that is currently being utilized by the New York State Department of Transportation for control efforts by maintenance crews. Of particular concern is the garlic mustard (*Alliaria officinalis*), whose 'infestations' can be detected 'before they become well established...thus maintaining a high quality natural landscape' (Spada and Brown, 2000).

To some observers, the notion of 'invasive' seems relative when applied to global movement of species, and from this perspective, the concept of 'exotic species' becomes an excellent example of an anthropocentric construction. However, as pointed out by Carson (1962), roadside vegetation is highly valued by vacationing tourists, whose 'good will' is a 'commodity prized by every chamber of commerce throughout the land' (p. 69). Aesthetic considerations are important as economic propositions. Land-use planners interested in promoting recreational use of the Adirondacks should be interested in Eurasian watermilfoil, garlic mustard, and other so-called exotic species that are invading the Adirondack Park. Native brook trout are more attractive to anglers than white suckers. A wetland of diverse vegetation is more appealing than one over-run with purple loosestrife. While planners might do what they can to curb anthropogenic sources of acid deposition and global warming, ecosystem management in the near future should focus on two questions: Which species are most damaging, aesthetically? Which species can be most efficaciously managed, at both least economic cost and smallest ecological side effect?

Planning Issues in the Adirondack Park Today

The previous section highlighted three of the most significant forces changing the Adirondack landscape: second-home development, with its attendant socio-economic effects; acid rain, with its ramifications for the natural world; and invasive species, with its implications for local tourism. Such forces of change have generated concern over how the park should be managed. The purpose of this section is to review controversial issues confronted by those who have taken a role in planning the future of the Adirondack Park.

What specific concern should be addressed by park planning? The answer varies depending on who answers the question. As discussed in the section before last, various people are apt to define the Adirondack Park differently. Consequently, people with different concerns emphasize distinctive issues when asked to describe problems that need to be addressed. As illustrated in the last section, for example, ANC focuses on the health of flora and fauna comprising natural communities within the Adirondack Park. This next section was written after interviewing representatives, reading documents, and consulting websites for several additional organizations interested in the future of the Adirondack Park. While admitting that other issues were worthy of consideration, each organization retains a specific focus, at odds yet compatible with all others.

Unit Management Plans and Open Space Acquisition for the Forest Preserve

The New York Department of Environmental Conservation (DEC) places clear emphasis on the preparation of Unit Management Plans (UMPs) for land in the Forest Preserve (2001). When the State Land Master Plan (Adirondack Park Agency, 1992) was enacted in 1972, each category of land (wilderness, canoe, primitive, wild forest, and intensive use) was composed of specific 'areas', (now called 'units'). The amount of land within each area (unit) ranged from a few acres to a few hundred-thousand acres. Table 8.3 delineates information on categories of state land, including number of areas (units) and total acres, both now and at the time that the State Land Master Plan was created.

Preparation of UMPs was required when the Adirondack Park Agency Act was signed into law in 1971. A UMP details the natural resources located within each unit, identifies possible recreational opportunities for that unit, and considers the ability of resources to provide for identified activities. UMPs are developed by staff at DEC in consultation with APA. Each UMP must be completed before additional recreational facilities can be constructed, such as trails, camping sites, parking areas, and boat launches. APA has final authority for ensuring that all UMPs are in compliance with the State Land Master Plan ('What is a Unit Management Plan?,' 2001). Despite legislative mandate, UMPs had not been prepared for over 100 units in the Adirondack Park by the late 1990s.

In October of 1999, the DEC initiated a major undertaking to complete all outstanding UMPs in the Adirondack Park over a five-year period. DEC appears to have a commitment to informing and involving the public; public meetings are periodically scheduled to fulfill that purpose. Seven informational sessions were held throughout the state in January of 2001. At that time, 22 UMPs had been completed, twelve were in progress, and about 80 were left to do within the five-year planning horizon (New York State Department of Environmental Conservation, 2001).

Open space acquisition by the State of New York has been a major concern for the Adirondack Council, perhaps the largest of the environmental lobbies operating within the Adirondack Park. Organized in 1975, the Adirondack Council is a private, non-profit organization with approximately 18,000 members. It claims to be 'dedicated to protecting the natural *and* human communities of the Adirondack Park through research, education, advocacy, and legal action' (emphasis added). Historically, however, the Council has demonstrated greater interest in the natural community. It has taken a lead role in the battle against acid deposition; for example, it established a popular program in which members of the public can 'retire' sulfur and nitrogen emissions by buying pollution allowance credits so they cannot be utilized by utilities. To protect biological habitats, the Adirondack Council advocates conversion of private land to Forest Preserve.

Table 8.3 Categories, areas (units), and acres for state land in the Adirondack Park

Category	Number of Areas/Units	Total Number of Acres* (1973)	(2000)
Wilderness	15	997,960	1,035,032
Canoe	1	18,100	17,016
Primitive	16	75,670	50,043
Wild Forest	15	undetermined	1,285,444
Intensive Use	70	undetermined	19,349
Administrative/Historic	--		1,944
Pending			38,537
		Total:	2,447,725

**Source*: 1973 data from *State Land Master Plan* (1973); 2000 data summarized from material compiled by Adirondack Park Agency (November, 2000).

The pace of land acquisition by New York State has quickened in recent years. In late 1997, the state purchased 14,700 acres of a private park (the Whitney Estate), which had long been coveted by proponents of open space. This addition to the Forest Preserve came at a cost of $13.9 million. Then, in 1999, the state acquired 29,000 acres in fee and 110,000 acres in easement from Champion Paper Company. This complicated negotiation was part of the largest public/private conservation agreement in United States history.

In total, Champion sold 325,000 acres in three states (New York, Vermont, New Hampshire) to the Conservation Fund of Arlington, Virginia, for $76 million. The Conservation Fund, in turn, sold title for 29,000 acres within the Adirondack Park to the State of New York, and title for 110,000 acres within the Adirondack Park to the Forestland Group LLC, a land management firm located in North Carolina. Forestland Group agreed to manage the land for logging only, using principles of sustainable forestry. The development rights on the land owned by Forestland were conveyed to the State of New York as a conservation easement. The total cost to the state was $29.4 million. Lands acquired in fee included 20 miles of river corridors, containing portions of the St. Regis, Deer, Oswegatchie, and Grasse Rivers (Adirondack Council, 1999).

The Adirondack portion of this agreement, of particular importance to the Adirondack Council, was made possible by both the 1996 Clean Water/Clean Air Bond Act and the 1998 New York State Open Space Conservation Plan. The Bond Act was proposed by Governor George Pataki and approved by New York voters in the second half of 1996. It provided $1.75 billion for environmental projects 'that serve to protect and restore the state's water, air, and natural resources' ('Clean Water/Clean Air Bond Act Report Shows Progress', 2001). $150 million was specifically earmarked for open space. Not only did the Bond Act furnish

$29.4 million for Champion lands in 1999, it also provided $13.9 million for acquisition of the Whitney estate in 1997.

The New York State Open Space Conservation Plan (OSCP) was authorized by a 1990 act of the state legislature. Not specifically focused on the Adirondack Park, the OSCP articulates a rationale and strategy for protecting open space and historical sites throughout the entire state. It also identifies specific areas that should be protected. The underlying framework is to consolidate relatively large areas, or 'blocks,' which are connected by linear corridors. The OSCP was initially enacted in November of 1992, but the authorizing legislation required that the original plan be updated every three years. Consequently, the OSCP was revised in 1995 and again in 1998. Each revision summarizes progress on implementation, incorporates additional desirable projects, and redefines priorities for future initiatives (New York Department of Environmental Conservation, 1998, p. iii). The OSCP works hand-in-hand with the Clean Water/Clean Air Bond Act. The first outlines what land should be acquired, and the second provides the financial resources to accomplish the task.

The Adirondack Council has played an active role in the OSCP. In this capacity, as well as others, the Council has lobbied diligently for protection of open space throughout the Adirondack Park. Most recently, the Council has focused on ways to improve conservation of private lands, especially in relation to practices of sustainable forestry. For example, the Council advocates a program of 'Green Certified Forestry' to promote and recognize sustainable practices on private lands (DiNunzio, 2001).

Growth and Development on Private Property

While the Adirondack Council focuses on enlarging the Forest Preserve, and DEC emphasizes UMPs for land already under state ownership, the Residents Committee to Protect the Adirondacks (RCPA) is concerned about development on private land within the park. Formed in 1990, RCPA is a private, non-profit organization with approximately 3,000 members. Its purpose sounds remarkably like that of the Adirondack Council. It is 'dedicated to the protection of the natural resources, resource-based economy, and rural communities of the Adirondack Park', and it 'pursues this agenda through grassroots organizing, advocacy, research, education, and legal action' (Bauer, 2001, p. 133).

Early in 2001, RCPA published a report on land-use regulation by both APA and local governments in the Adirondack Park. The report concluded that APA has been somewhat successful, since recent growth has occurred in areas zoned for highest densities (hamlet, low intensity, and moderate intensity). RCPA also found that development has been heaviest along roadsides and lakeshores. Building along highways reduces the feeling of open space in the park. The reason for this situation is the failure of APA to adopt a 'clustering' policy to limit sprawl and make it less visible, especially in areas of rural use. Building on lakeshores is problematic because these areas are among the most fragile ecosystems located within the Blue Line. The root of this problem lies within the LUDP. Many lakeshores are zoned as moderate intensity, with a minimum density guideline of only 1.3 acres per dwelling and a minimum shoreline width of only 100 feet per lot (Bauer, 2001, p. 7).

RCPA calculated that local governments throughout the park permitted 47,762 activities between 1990 and 1999. Many were for small projects, such as renovations and expansions to existing buildings. Of the total activities, 8,589 were for new residential, commercial, and industrial structures. Many authorizations for new buildings were in hamlets, where APA has no regulatory jurisdiction. During this same period (1990-1999), APA permitted 9,647 activities. After a permit is issued by APA, a permit must also be obtained from the local town or village before construction may commence; in other words, the 47,762 activities approved by local governments include the 9,647 authorized by APA, or approximately 20 per cent of the total. APA approved 3,731 new residential, commercial, and industrial structures, or 43 per cent of the total new buildings between 1990 and 1999. In recent years, then, the majority of both total and new development in the Adirondack Park has been regulated by local governments (Bauer, 2001, p. 6).

Given that a preponderance of land-use is decided at the local level, it seems reasonable to ask: how effective is land-use planning by the 92 towns and 11 villages in the park? Approximately 60 per cent of these local governments have comprehensive master plans, but the quality of these plans varies widely. Only 14 towns and one village have plans approved by APA. It was anticipated, when the LUDP was enacted in 1973, that many local governments would complete master plans that could be approved by APA. The opportunity to gain authority over projects without regional significance (class B projects) was ostensibly the incentive for towns and villages to carry APA approval. For the 14 towns and one village with APA-approved plans, the regulatory function appears to have been transferred effectively. During the decade of the 1990s, APA reviewed just eight per cent of total development and 20 per cent of new structures in these 15 jurisdictions, compared to 20 per cent and 43 per cent, respectively, for the park as a whole (Bauer, 2001, pp. 30-31). For the remaining 88 towns and villages in the Adirondack Park, APA is still shouldering more of the regulatory burden than framers of the LUDP had originally envisioned.

The RCPA study also revealed that 60 of the 103 towns and villages in the Adirondack Park have adopted zoning ordinances and 61 have enacted subdivision regulations (Bauer, 2001, p. 7). However, some laws are not very stringent, and other local governments are simply refusing to coordinate their land-use planning efforts with APA. If things are to work better in the future, APA needs to provide financial incentives to towns and villages within the park. The state underwrote such a program of incentives, called the Adirondack Park Local Planning Assistance Program, soon after the APA was established in 1971. The idea was that monetary grants would more likely produce better planning, as well as augment the amount of cooperation between APA and local governments. Unfortunately, the program was terminated in the 1980s. RCPA would like it restored, with both an annual budget of one million dollars and a team of professional circuit riders, who could provide training for officials of towns and villages, including code enforcement officers (Bauer, 2001, p. 79). The Adirondack Council agrees on the need for circuit riders, citing a 1994 report by a special task force appointed to streamline and improve APA procedures (recommendation 99 in *The Citizens Task Force Report,* cited in The Adirondack Council, 2001, pp. 6-7).

From the perspective of RCPA, time is of the essence. As of 1999, there were 83,475 structures within the Adirondack Park, of which approximately 10 per cent (8,589) had been built in the 1990s (Bauer, 2001, p. 5). The potential for future development within the park is massive, as illustrated by build-out calculations, such as those in table 8.4. RCPA summarizes the issue in the following way:

> How well the State, in partnership with Adirondack towns and villages, handles and regulates growth remains the most pressing question for the future of the Adirondack Park. The Adirondack Park is an attractive place for people to move to, buy vacation property in, and return to because it is a landscape like no other in the U.S. Though just a half-day's drive from the major population and economic centers of the eastern United States, the Adirondack Park is still in many respects a wild landscape dominated by big open spaces dotted with small towns. But residential growth gone wrong will kill the goose that lays the golden egg. The Park's incredible array of water – lakes, rivers, ponds, streams – its immense, deep forests and spectacular mountains underwrite the Adirondack economy, the character of Adirondack communities, the quality of life for residents, and the Adirondack experience. The Adirondack Park offers a wild landscape, something in direct contrast to and long lost in the rest of eastern U.S. But if growth continues at the same rate and the same patterns over the next 25 years as it has over the past 25, the wild landscape that Adirondack residents and visitors enjoy today may be lost (Bauer, 2001, pp. 78-9).

Tax Base and Socio-Economic Conditions for Permanent Residents

While RCPA is concerned about the capacity of APA and local governments to undertake planning that will protect the character of the Adirondack Park, the Adirondack Association of Towns and Villages (AATV) focuses on the ability of local governments to provide public services and to improve socio-economic conditions for park residents. Formed in 1992, the membership of AATV includes elected officials from the 103 towns and villages that are partly or entirely located within the Blue Line. The purpose of AATV is 'to act as a representative of the towns and villages of the Adirondacks in addressing issues unique to local government and residents within the Adirondack Park'. Furthermore, it is the position of AATV to 'provide a forum for Adirondack town and village officials to receive and exchange relevant information from both the public and the private sector' [and] 'take all necessary and proper actions to preserve strong and effective town and village government in the Adirondacks' ('AATV Primary Positions', 2001).

Table 8.4 Categories, acres, and maximum build-out for private land in the Adirondack Park

Category	Number of Acres* (2000)	Intensity Guideline (acres/dwelling)	Maximum Build-Out (# of dwelling)
Hamlet	53,556	--	unlimited
Moderate Intensity	102,096	1.3	78,535
Low Intensity	270,446	3.2	84,514
Rural Use	1,017,395	8.5	119,694
Resource Management	1,582,528	42.7	37,062
Industrial Use	12,290	--	
		Total:	3,038,311

**Source*: Summarized from material complied by Adirondack Park Agency (November, 2000).

Within the Adirondack Park, government accounts for the largest share of employment, with more than 30 per cent of the total number of jobs. The service sector is second, with slightly less than 30 per cent. In terms of salary, more than 40 per cent is derived from government sources, strikingly more than 16.4 per cent for the entire United States. Although inflation-adjusted payrolls increased in total by more than seven per cent between 1992 and 1997, annual salaries for permanent residents were still considerably below both state and national averages (Beideck, 1999).

Many residents are eager to enhance employment opportunities through expansion of tourism. As documented earlier, permanent residents have relied on economic benefits from tourism since the decline of agriculture in the region 100 years ago. A 1996 survey by the Lake Placid/Essex County Visitors Bureau estimated that 8.5 to 10 million overnight visits are made to the Adirondack Park each year. Each visit brings at least $100 into local economies. Specific communities are looking at different means to increase tourism. Lake Placid, for example, promotes its competitive sports facilities; Indian Lake advertises snowmobiling; and various towns along Lake Champlain focus on historical interpretation and bicycling (Brooks, 1999).

AATV is particularly interested in the economic benefits of attracting snowmobiles ('AATV 2001 Legislative Agenda,' 2001). With few exceptions, notably in the vicinity of a handful of ski areas, the number of visitors in winter is much less than at other times of the year. Unemployment climbs during this season, which is typically quite long. Winter in most of the Adirondack Park begins with the onset of November and ends in mid-April, or later. In the opinion of AATV, snowmobiles can boost the local economy substantially at a time when the need is greatest.

George Canon, president of AATV, also stated in a recent interview that 'the Adirondack Association of Towns and Villages consistently expresses their opposition to the state acquisition of land in fee. Some of our resolutions, almost annually, provide this issue on the legislative agenda'. The problem lies in the fact that when the state buys land in fee, it prevents forestry from being a major employer. This leaves recreation and tourism, but according to Canon, some visitors contribute more than others. He believes that non-motorized recreationists, for the most part, do not contribute much to the local economy; many drive into the park in the morning, carry their own food, and go home at the end of the day (Canon, 2001).

AATV is also concerned about revenue for the provision of public services by local governments. One issue is the size of the tax base. With acquisitions for the Forest Preserve, the amount of private land in some towns has diminished significantly. However, the state of New York compensates local governments in relation to the amount of land in state ownership. Canon believes that New York State is a 'decent taxpayer' since payments in lieu of taxes are similar to taxes paid by private landowners. In other words, the formula for calculating reimbursement is fair and the size of monetary allocations is sufficient. The problem in relation to the tax base is preferential assessment for private land in forestry.

The program for preferential assessment of timberland in New York is known as the Fisher Forest Tax law. It provides that '[l]ands devoted to growth of forest products should be assessed at a level which recognizes this use rather than at a level reflecting devotion of the land to another purpose' (New York Department of Environmental Conservation, n.d.) The legislation requires a minimum of 25 acres be devoted to forest practice, and that land remain in "forest crop production" for a minimum of eight years. Removal of land from forestry is subject to a tax rollback of five years by each taxing jurisdiction.

Use-value assessment is established by the New York State Board of Equalization and Assessment, and this value must be used by local assessors in determining taxes for eligible lands. By taxing land at use-value rather than market-value, local governments receive less revenue. What AATV would like is for the State of New York to reimburse taxing jurisdictions in a manner similar to compensation for land in the Forest Preserve.

Another issue in relation to revenue is state assistance for local code enforcement. For many years, the State of New York directed a portion of fees collected from insurance companies to local governments for code enforcement. Beginning in 1992, this money was instead diverted to the general fund of the state, thereby eliminating a source of funding for implementing local police powers. AATV would like this program re-instated. In its opinion, 'elimination has dealt a devastating blow to local code enforcement efforts and many programs have been dismantled, putting the residents at risk' (AATV 2001b).

Financial Resources for Land-Use Planning

Despite variations in why they want to plan, all of the above organizations have expressed a desire for financial resources to undertake planning. The DEC needs funding to complete UMPs; the Adirondack Council advocates additions to the Forest Preserve under OSCP, which requires more money. RCPA would like restoration of the Local Planning Assistance Program; and AATV is lobbying to restore state assistance for local code enforcement. The fundamental goals and objectives of each group vary, but a thread of common concern unites them.

APA is in complete agreement. James Hotaling, Principal Adirondack Park Local Planning Assistance Specialist, said in a recent interview: 'as we get further along, planning boards are called upon to do more and more technical activity, such as analyze traffic and soil, and to look at the land. So, in the future we will continue to need to have those skills developed and maintained by planning board members' (Hotaling, 2001).

The problem is part of an interconnected set of issues. As documented by RCPA, local land-use planning has not adequately assumed a larger share of total planning activity. If towns and villages are to undertake really effective planning, much of the work will become increasingly technical. As noted by Hotaling, more training will be required for local planners, and as stated by AATV, resources are already needed for local code enforcement. At the same time, the amount of development in the last decade appears to be growing, not shrinking. Thus, until the time local governments engage 'approved' planning, APA has a greater amount of work to do.

Another critical issue outlined in a recent report by the Adirondack Council, is enforcement of LUDP regulations by APA staff. The number of possible violations of the LUDP grows at a rate faster that the number of cases that are settled or otherwise resolved. In 1999-2000, 411 new cases were opened, while only 178 cases were closed. The total backlog estimated by APA staff is between 2000 and 3000 cases (Adirondack Council, 2001, pp. 4-5). APA staff is clearly unable to keep up, and some of the fallout is certainly disturbing. One troubling scenario is the issuance of 'after-the-fact permits'. According to the Adirondack Council, '[a]fter-the-fact permits routinely sanction out-of-compliance construction' (Adirondack Council, 2001, p. 3).

Not only do backlogged cases go unresolved, but APA is unable to seek out violators who may not be adhering to the LUDP or conditions attached to their permits. The two most significant sources of information regarding violations are complaints from neighbors and applications for new permits which, when reviewed, reveal prior violations. APA staff rarely has time to conduct its own investigations of compliance. One of the specific recommendations in the report of the Adirondack Council is to increase the number of enforcement officers, from three to six, and to double the number of attorneys who work on enforcement cases, also from three to six. In fact, the most recent New York State budget allowed APA to hire one additional enforcement officer and one more attorney. Another suggestion by the Adirondack Council is that APA be given the authority to collect (1) fees from permit applications, as a means to defray costs of both processing its review and checking compliance, and (2) fines levied against violations, so these monies can be returned directly to the Agency rather than the general fund of the state (Adirondack Council, 2001, pp. 9-10). Both suggestions would seem imminently reasonable to those with an interest in the future of the Adirondack Park.

Conclusion

Looking ahead, long-range policy must accommodate the matrix of public and private property in the Adirondack Park. The Forest Preserve will take on increasing importance as a natural area. Conservation biologists have already identified the Adirondack region as a core or block of habitat that might be connected by wildlife corridors (Medeiros, 1992/93). As well, permanent residents will live in the park for the foreseeable future, and they will want to improve the quality of their socio-economic lives. Development will doubtless continue on private land. Given abundant natural resources on land owned by both timber companies and the state, the Adirondack Park represents a tremendous opportunity to develop a model for economic sustainability. Land-use planning in the future will need to place economic growth within natural resource limitations (Thorndike, 1997).

People throughout the Adirondack Park share a distinct culture that makes their lives unique. Residents must live where the state owns approximately half the land and where the LUDP constitutes a level of regulation that most others in rural America have never encountered. They are also aware that the quality of life they enjoy is threatened by a host of problems. The economy remains precarious, and young people are often unable to find employment in the communities where they were raised. The ongoing objective is to improve economic conditions without seriously altering the wild that makes the Adirondack Park special (Terrie, 1997, pp. 181-182).

The natural world of the Adirondack region has always been significant for economic purposes, whether for farming, logging, and transportation in the nineteenth century, or recreation and tourism in the twentieth century. Catering to visitors is likely to be the primary source of income for park residents in the future. The price of economic vigor for year-round residents of the Adirondack Park is a healthy natural environment, since open space and biological diversity are necessary to encourage recreation and tourism.

The risk of an economic strategy based on recreation and tourism is that it will only exacerbate differences in the way permanent residents and visitors perceive and define the Adirondack Park. A larger number of people will be attracted to the park to have more fun for a greater proportion of the year. A larger number of people will be employed in the park at higher wages for a greater proportion of the year. Without mutual understanding, land-use planning and other measures of public policy will remain controversial. One of the particular difficulties for planners will be to convince the local population that roughly nine million outsiders view the park as a playground, and to persuade these visitors, as well as the 110,000 seasonal residents who own second homes, that the park is home and a source of livelihood for 130,000 permanent residents.

In sum, the future must solve circumstances created by the past. When the Blue Line was drawn around scattered parcels of Forest Preserve, park proponents envisioned that the state could and would eventually acquire all private land therein. This thinking was over-ambitious. As time unfolded, park proponents succeeded in defining an image of the Adirondack landscape, but their actions on a practical level were less successful. Much private land was indeed acquired by the state. However, much remained in private ownership. As more and more visitors were attracted to an image of an 'Adirondack Park', private land came under greater and greater development pressure.

Land-use planning in the Adirondack Park has been controversial because residents who have lived within the park perceive and define the landscape differently than both those who have framed protective legislation and those who visit. Sometimes one definition prevails, as when APA and the LUDP was enacted; on different occasions the other perspective prevails, as when the recommendations of the Cuomo Commission were largely shelved. These definitions are embodied, to varying degrees, in the many organizations that represent special interests throughout the Adirondack Park. Each organization has its own set of priorities, which reflect how its members define the park and which give the organization its own identity. Yet, despite these distinctions, there is common concern that sufficient financial resources be available for planning the future of the park. Starting from this point of mutual agreement, it is possible that land-use planning in the future can overcome discontent created by efforts to plan in the past. By understanding when and how the seeds of discontent were sown, planners might yet help the Adirondack Park yield a sustainable future.

Acknowledgements

Appreciation is expressed to the following individuals who provided information and offered opinions about the perspective of their organizations: Peter Bauer of the Residents Committee to Protect the Adirondacks, Bill Brown of the Adirondack Nature Conservancy, George Canon of the Adirondack Association of Towns and Villages, Michael DeNunzio of the Adirondack Council, Jim Hotaling of the Adirondack Park Agency, Victoria Hristovski of the Adirondack Park Agency, Nina Schoch of the Adirondack Nature Conservancy, and Todd Thomas of the Residents Committee to Protect the Adirondacks. At St. Lawrence University, Joan Larsen of the Owen D. Young Library assisted in gathering statistical information, and Nancy Alessi of the Environmental Studies Program helped to prepare information for publication. Special thanks to Carrie Johns of the Environmental Studies Program for guidance and feedback, especially in the section on acid deposition.

References

'AATV 2001 Legislative Agenda', http://www.aatvny.org/Legislative%20Initiative.html, May 5, 2001.

'AATV Primary Positions', http://www.aatvny.org/primary_positions.html, May 5, 2001.

Abbey, Edward (1968), *Desert Solitaire: A Season in the Wilderness*, Ballantine Books, New York.

A Citizen's Guide to Adirondack Park Agency Land Use Regulations, booklet available from Adirondack Park Agency, P.O. Box 99, Ray Brook, NY 12977.

Adirondack Council Newsletter, January, 1999), '139,000 Acres of Park Protected'.

Adirondack Park Agency (1972), *State Land Master Plan*, in consultation with the Department of Environmental Conservation.

Adirondack Park Agency (1973), *Land Use and Development Plan.*

Adirondack Park Agency, *Summary of Adirondack Park Agency Authority over Land Use and Development and Subdivisions*, chart available from Adirondack Park Agency, P.O. Box 99, Ray Brook, NY 12977.

Analysis of Rates and Patterns of Development (North Creek, The Residents' Committee to Protect the Adirondacks, New York.

Beideck, Alan (1999), 'Blue Line Labor: An Update of Employment and Payrolls in the Adirondack Park', *Adirondack Journal of Environmental Studies*, Vol. 6, No. 2, pp. 18-22.

Brooks, Cali (1999), 'Perspectives on an Adirondack Landscape', *Adirondack Journal of Environmental Studies*, Vol. 6, No. 2, pp. 11-17.

Burroughs, John (1889), 'Adirondack', in *Wake Robin*, Houghton Mifflin Company, Boston.

Canon, George, Personal Communication, May 10, 2001.

Carson, Rachel (1962), *Silent Spring*, Houghton Mifflin Company, Boston.

'Clean Water/Clean Air Bond Act Report Shows Progress', http://www.dec.state.ny.us/website/press/pressrel/99-109.html, April 24, 2001.

Cobb, Thomas, L. (2000), 'On the 1967 Proposal for an Adirondack Mountains National Park', *Adirondack Journal of Environmental Studies,* Vol. 7, No. 1, pp. 10-12.

Colvin, Verplanck (1871), 'Ascent of Mount Seward and Its Barometrical Measurement', *New York Senate Documents No. 68*, Albany.

Cook, Marc (1881), *The Wilderness Cure*, William Wood and Company, New York.

Craig, B.W. and Friedland, Andrew J., (1991), 'Spatial Patterns in Forest Composition and Standing Dead Red Spruce in Mountain Forests of the Adirondacks and North Appalachia', *Environmental Monitoring and Assessment*, Vol. 18, pp. 129-140.

DeHayes, Donald H., Schaberg, Paul G., Hawley, Gary J., and Strimbeck, G. Richard (1999), 'Acid Rain Impacts on Calcium Nutrition and Forest Health', *Bioscience*, Vol. 49, No. 10, pp. 789-800.

Department of Environmental Conservation, *New York Forest Tax Law,* pamphlet, 316 Washington St., Watertown, NY 13601.

Department of Environmental Conservation and the Office of Parks, Recreation and Historic Preservation (1998), *Conserving Open Space in New York State.*

DiNunzio, Michael, Personal Communication, January 29, 2001.

Donaldson, Alfred A. (1921), *A History of the Adirondacks,* 2 vols., Century Books, New York.

Driscoll, Charles T., Lawrence, Gregory B., Bulger, Arthur J., Butler, Thomas J., Cronan, Christopher S., Eager, Christopher, Lambert, Kathleen F., Likens, Gene E., Stoddard, John L., and Weathers, Kathleen C., (2001), 'Acidic Deposition in the Northeastern

United States: Sources and Inputs, Ecosystem Effects, and Management Strategies', *Bioscience*, Vol. 51, No. 3, pp. 180-198.
Environmental Protection Agency (1999), *Progress Report on the EPA Acid Rain Program*, Report EPA-R-99-011, Washington, U.S.
Gallos, Philip, L. (1985), *Cure Cottages of Saranac Lake*, Historic Saranac Lake, Saranac Lake, NY.
Graham, Frank, Jr. (1978), *The Adirondack Park: A Political History*, Alfred A. Knopf, New York.
Hammond, Samuel H. (1857), *Wild Northern Scenes; or, Sporting Adventures with the Rifle and Rod*, Derby and Jackson, New York.
Harris, Glenn (1999), 'Nineteenth Century Agriculture on Lands within the Present Blue Line of the Adirondack Park', presented at the Sixth Annual Conference on the Adirondacks, Saranac Lake, May 27.
Harris, Glenn, Hornaday, Krista, and Trumble, Ralph (2000), *Agricultural Activities in the Journals of Charles Stafford, Essex NY, 1856-1891*, Community Service Report submitted to the Essex Community Heritage Organization, St. Lawrence University Environmental Studies Program, Canton, NY.
Harris, Glenn and Wilson, Seth (1993), 'Water Pollution in the Adirondack Mountains: Scientific Research and Governmental Response, 1890-1930', *Environmental History Review*, Vol. 17, No. 4, pp. 47-71.
Headley, Joel Tyler (1849), *The Adirondack; or, Life in the Woods*, Baker and Scribner, New York.
Hotaling, James, Personal Communications, February 2 and May 10, 2001.
Kaiser, Harvey H. (1982), *Great Camps of the Adirondacks*, David R. Godine, Boston.
Krajick, Kevin (2001), 'Long-Term Data Show Lingering Effects from Acid Rain', *Science*, Vol. 292, pp. 195-196.
Kretser, Heidi (2000), 'Empty Spaces Offer a World of Opportunity', *Adirondack Journal of Environmental Studies*, Vol. 7, No. 1, pp. 7-9.
Kudish, Michael (1995), 'Old-growth Forests in the Catskills and Adirondacks', *Wild Earth*, Vol. 5, No. 3, pp. 52-53.
Leopold, Aldo (1966), *A Sand County Almanac*, Sierra Club/Ballantine Books, New York.
Marsh, George P. (1864), *Man and Nature; or, Physical Geography as Modified by Human Action*, Charles Scribner and Co., New York.
McCormick, John (1997), *Acid Rain: The Politics of Acid Pollution*, 3rd ed., Earthscan Publications, London.
McHarg, Ian (1969), *Design with Nature*, Doubleday and Company, Garden City, NJ.
McMartin, Barbara (1992), *Hides, Hemlocks and Adirondack History*, North Country Books, Utica.
McMurry, Sally (1999), 'Evolution of a Landscape: From Farm to Forest in the Adirondack Region, 1857-1894', *New York History*, Vol. 80, No. 2, pp. 117-152.
Medeiros, Paul (1992/93), 'A Proposal For An Adirondack Primeval', *Wild Earth*, Special Issue, pp. 32-42.
Mission Statement', http://nature.org/states/newyork/Adirondacks/about/, May 5, 2001.
Murray, William H. H. (1869), *Adventures in the Wilderness; or, Camp Life in the Adirondacks*, Fields and Osgood, Boston.
Nash, Roderick (1982), *Wilderness and the American Mind*, 3rd ed., Yale University Press, New Haven.
National Atmospheric Deposition Program (2000), *National Atmospheric Deposition Program 1999 Wet Deposition*, NADP Data Report 2000-02, Illinois State Water Survey, Champaign, IL.

Nelson A. Rockefeller Institute of Government (2000), *2000 New York Statistical Yearbook*, Hamilton Printing, Rennselaer, NY.

New York State Department of Environmental Conservation *(1997), Common Nuisance Aquatic Plants in New York State*, prepared by the Lake Services Section.

Park, Chris C. (1987), *Acid Rain: Rhetoric and Reality*, Methuen, New York.

Pasquarello, Thomas, Buerger, Robert, and Randorf, Gary (1994), 'Wilderness and the Working Landscape', *Adirondack Journal of Environmental Studies,* Vol. 1, No. 1, pp. 19-23.

Roy, Karen, Kretser, Walter, Simonin, Howard, and Bennett, Edward (2000), 'Acid Rain in the Adirondacks: A Time of Change!', *Adirondack Journal of Environmental Studies,* Vol. 7. No. 1, pp. 26-32.

Schoch, Nina (April 2, 2001), Personal Communication.

Sheehan, John F. (1996), 'Acid Rain Still a Scourge in Adirondacks', *Wild Earth*, Vol. 6, No. 1, pp. 34-36.

Shortle, W.C., Smith, K.T., Minocha, R., Lawrence, G.B., and David M.B. (1997), 'Acid Deposition, Cation Mobilization, and Stress in Healthy Red Spruce Trees', *Journal of Environmental Quality*, Vol. 26, pp. 871-876.

Spada, Daniel and Brown, William (May 25, 2000), 'Discussion Forum: Invasive Exotic Plant Species in the Adirondack Park – The State of Our Knowledge', Seventh Annual Conference on the Adirondacks, Saranac Lake.

Stager, Curt (1996), 'Update on the Ecological Condition of Adirondack Lakes', *Wild Earth*, Vol. 6, No. 1, pp. 29-33.

Task Force on Expediting Adirondack Park Agency Operations and Simplifying its Procedures (1994), *The Citizens Task Force Report.*

Temporary Study Commission on the Future of the Adirondacks (1970), *The Future of the Adirondack Park*, (reprint, The Adirondack Museum, 1971, Blue Mountain Lake, NY. Terrie, Philip G. (1997), *Contested Terrain: A New History of Nature and People in the Adirondacks*, The Adirondack Museum/Syracuse University Press, Blue Mountain Lake, NY.

Terrie, Philip G. (1985), *Forever Wild: Environmental Aesthetics and the Adirondack Forest Preserve,* Temple University Press, Philadelphia.

The Adirondack Council *(2001), Falling Further Behind: The Truth About Environmental Enforcement in the Adirondack Park,* Elizabethtown, NY.

The [Adirondack Nature Conservancy] Flicker (Winter, 1993), 'Partnership Protects Species and Jobs'.

The Commission on the Adirondacks in the Twenty-First Century (1990) report, *The Adirondack Park in the Twenty-First Century.*

Thorndike, Elizabeth (1997), 'The Adirondack Park in the 21st Century', *Adirondack Journal of Environmental Studies*, Vol. 4, No. 2, pp. 14-17, 30-32.

Todd, John (1983), *Long Lake*, 1845, (reprint, Harbor Hill Books, Harrison, NY, 1983).

'Unit Management Plans for State Land' (April 24, 2001), http://www.dec.state.ny.us/website/dlf/publands/ump/index.html.

Watson, Winslow C. (1852), 'A General View and Agricultural Survey of the County of Essex', *Transactions of the New York State Agricultural Society,* (reprint, Van Benthuysen, Albany, 1853).

'What is a Unit Management Plan?' (April 24, 2001), http://www.dec.state.ny.us/website/dlf/publands/ump/umplans.html.

Chapter 9

Lessons For The Future

Owen J. Furuseth and Mark B. Lapping

The North American experience with regionalism and regionally structured planning frameworks has been mixed. On one level, we have theoretical grounding and empirical evidence that strongly suggests that regional approaches to the problems of environmental management, economic development, and social equity can offer solutions superior to locally centered schemes. However, because regional approaches imply both fewer opportunities for citizen input and fewer safeguards for equitable decision-making, at least from the local perspective, the recurring challenge is how to achieve some degree of fairness, overcome suspicions, align social and economic needs with environmental priorities, and build successful regional planning strategies.

Each of the chapters in this volume addresses the challenges facing planners and citizens grappling with finding a balance between short-term human needs and the longer term sustainability of the natural resource base. The roots of regionalism and regional planning that capture the mutual importance of the human world and the natural environment were fashioned in the 1920s and 1930s by Lewis Mumford, Benton MacKaye, Catherine Bauer, Clarence Stein and others involved with the Regional Planning Association of America (RPAA). The integrated approach to regional development sought balance between social needs and the natural ecosystem. Without using the terms 'sustainability' or 'smart growth', the RPAA approach to planning was remarkably progressive and visionary. Theirs was an approach to regional planning that advocated adapting human form and livelihood to the critical forces in the regional ecosystem, recontextualizing cities in relation to nature, and building economics based upon natural advantages of a region and avoiding over specialized regional economies and economic frameworks built on systems of commodification (Luccarelli, 1995, p.3). In particular, Mumford's provident critiques of American planning are precursors for contemporary advocates for responsible growth and development. In turn, Mumford and MacKaye's (1962) notions for regional planning find resonance and validation in the experiences presented by the authors in this volume.

Cumulatively, this volume presents an insightful discourse on the land and resource challenges facing elected officials, resource managers and community leaders in some of North America's most unique and environmentally significant rural regions. The presentations are diverse in their background, positions and philosophies on moving toward new ways to accommodate growth and change from a perspective of sustainability. The individual policy regimes are as varied as the geophysical settings of these special places.

Nonetheless, a number of common themes are woven into all of the chapters.

Thus, if we are not presented with the answers to the myriad of problems concerning governance and management of rural resources, we are offered compelling evidence of strategies and organizational forms that can be used to break the impasse or loosen the policy gridlock that often surrounds these conundrums. The contributors do not oversimplify the problems of crafting land use and resource plans in societies where the primacy of private property rights makes the legitimacy of the public interests a controversial and often contested notion.

In a recent policy report for the Lincoln Institute of Land Policy, Kathryn A. Foster (2001) examined the challenges facing regionally structured governance systems in the United States. While her analysis was largely focused on the urban-suburban regional framework, Foster's assessment of the impediments confronting regional government entities has saliency to the problems and issues facing rural regional planning and management. Foster's special challenges included:

- overcoming a weak sense of regional identity;
- finding consensus on strategies for regional change;
- securing the benefits of a 'big tent' coalition without succumbing to the fragility of diverse alliances;
- overcoming a strategic bias toward relatively uncontentious issues of economic development and away from more controversial equity and land use issues; and
- responding to often inconsistent federal and state policies.

Much of the struggles surrounding the development and implementation of regional planning efforts in New Jersey's Pinelands and New York's Adirondack region can be linked back to questions of regional identity. As Robert Mason and Glen Harris and Michael Jarvis lay out the conflicts between traditional, if you will 'native' residents, versus supra-local interest groups and newer residents, the questions of whose voices represent the real interest of the region and whose values are enshrined in the planning policies and growth management programs are critical. In this context, the absence of regional identity or, alternatively, loosely held concepts of place uniqueness or boundaries weakens the legitimacy of the regional efforts and requires stronger efforts to address, if possible, oppositional interests.

In a similar fashion, the process of finding consensus for regional planning in rural regions is often fraught with navigating through ideological and cultural values that create barriers for consensus-building. Whether working to develop land use planning frameworks for rural New England in Maine and Vermont or devising environmentally sustainable land use regulations and conservation strategies for fragile resources in Ontario, California, Nevada, New Jersey and Florida, planners and resource managers are challenged to find ways to build trust, foster a sense of fairness, and offer solutions that provide accountability and stability for all the affected parties. In some cases, as we have seen in this book, constitutional questions lead to the involvement of the judiciary in rendering a consensus. Clearly, a top-down, forced consensus, whether it comes from a judge or a legislature or executive action, can create confusion and a short-lived and thin consensus. In that way, consensus derived from open public discussion building work more effectively and provide more satisfactory answers to regional resources and land use issues.

Coalition building in the regional planning arena, and especially when it involves distributional decisions in environmental or resource management, is rarely a simple process. Different strategies with alternative mechanisms to achieve compromise and consensus need to be crafted for individual settings. What worked in Vermont clearly would not garner much success in the Adirondack Park. Most importantly, coalitions are always at some risk of collapsing. As seen in Maine and New Jersey, a broadly based coalition can come apart under the weight of the diverse interests that compose it. All too often, in the aftermath of coalition building, constituent interests groups may feel that they were bullied or bluffed into participating and thus taken advantage of. This remorse, in turn, can mean the end of a successful consensus and ongoing conflict with resource planning as its arena.

In an environment where decision-making is hobbled by issues of long-term public support and inadequate institutional support structures, there is a rational reason to focus one's attention on broadly achieved 'easy' policy outcomes. For planners and decision-makers in rural regions, the first order of business may be the creation of structures and strategies for the most popular issues. Where there is fiscal largesse, as in the Everglades, it is easier to address property right issues. Similarly, if restrictions of land subdivision disproportionally affect out-of-staters, as in Vermont and Maine, then the rules and regulations surrounding land development are easily more palatable and offer fewer risks to long-term planning. But serious regional planning and land management cannot avoid the contentious questions, and these efforts will inevitably create 'winners and losers' and disputes are inherent. The challenge is to move forward on the more confrontational issues in a way that avoids stalemate, minimizes conflict, and maximizes long-term policy success.

One of the problems facing rural regional planners are the mixed messages and inconsistencies in policies and programs from more senior federal or state/provincial partners. As Hugh Gayler recounts, many of the long-standing issues facing the land conservation interests in the Niagara region were directly the result of weak and vacillating messages from the Ontario government. This has been true, to a lesser degree, in both the Everglades and the Lake Tahoe basin where different federal political regimes have likewise been inconsistent. While it may be beyond the control of regional planning advocates to eliminate these mixed signals, planning frameworks need to be constructed so that they are robust enough to respond to the countervailing messages from beyond the region.

As seen in this volume, the creation of effective and sustainable regional planning structures for rural areas remains an ongoing challenge for concerned citizens and leaders. As might be expected, advocates for regionalism and reform are constantly facing resistance to attempts to shift responsibility for decision-making to regional-oriented governance and decision-making entities. Unfortunately, regionalism is most easily recognized in times of crisis or when the overt failure of local decision-making is most blatant. Shortages in water, deteriorated water quality, traffic congestion, visual pollution, and skyrocketing local taxes are all triggering factors for public dissatisfaction with the status quo. However, the enduring quality of land use decisions or the irreversible degradation of the environment may render progressive regional land use planning and policies too late.

What summative lessons can we apply from the commentaries presented in this text for rural regions in other parts of North America? Vision is critical. At the core of all successful rural regional plans is an articulated and projected vision around which key stakeholders can rally. This vision should reflect a grounding in the realities of human and environmental parameters, while also projecting an alternative future that can meet regional challenges and build the capacity of the region for effective action.

The regional vision is at the nexus of creating processes and products to inform consensus building, regional problem-solving, and the creation of mechanisms and resources to address key issues and problems. To be sure, the articulation of a vision will not ensure the immediate success of the regional initiative. Months and even years may pass before effective regional capacities are in place. But without a strong vision, the task is flawed.

At a time when powerful supra-regional forces such as economic globalization and social restructuring have shaken traditional assumptions and the 'social contracts' that govern rural life, new planning regimes built upon a changing ethos are increasingly viable. As we noted at the beginning of this volume, the past forty years has witnessed a maturation in the thinking about planning frameworks, public interest, and environmental needs (Lapping and Furuseth, 2003). While institutional and governance reforms have been marked by 'starts and stops,' social and economic changes have, in many ways, constructed new contexts where the values and importance of place are supplanting rural suspicions and inertia. Is there any doubt that the progressive land use planning or resource management programs presented in this volume would have been possible were it not for cascading effects of economic and social change that swept North America during the latter portion of the twentieth century?

The key point for rural planners is how to tap into the changing ethos. In the case studies presented herein, planners were able to shape new policies and tools in conjunction with a fundamental rethinking of traditional norms. Indeed, where planning processes were successfully implemented there is a shared purpose between plans and the evolving attitudes and thinking about the rural region.

While the North American experience with rural regional planning has been poorly developed when compared with other Western, developed areas, the timing may now be right. Changing social, economic, and environmental processes have intertwined in rural North America in the recent past in new ways, challenging the status quo, tearing down comfortable norms, while exacerbating problems. The 'trickle down' from urban to rural, once thought to be positive and benign, have turned questionable. In the face of changed political economies, there is a growing urgency for rural localities to reexamine the way to development and use local resources. Exploitative processes pale when time scale is adjusted and distributional dimensions are considered.

Regional-based and environmental sustainable planning frameworks offer constructive alternatives to the traditional development model, while helping to buffer rural communities from the disruptions afflicted on localities by globalization. Ironically, in the current political environment, regional rural planning regimes may provide more affirmation for local community structures than the traditional governance arrangements. Admittedly, the stories told in this book reflect the experiences from many of North America's most environmentally unique rural regions. Thus, we would expect an elevated level of concern for environmental stewardship. Nonetheless, the experiences and processes at work in Vermont, Ontario, Florida and the other 'Big Places' are not isolated. Rather across the North American countryside, the need and concern for better, more sustainable planning is increasingly evidenced.

References

Foster, Kathryn A. (2001), *Regionalism on Purpose*, Lincoln Institute, Boston.

Lapping, Mark B. and Furuseth, Owen J. (2004), 'Introduction,' in Mark B. Lapping and Owen J. Furuseth (eds.), *Big Places, Big Plans*, Ashgate, Burlington, pp. 1-3.

Luccarelli, Mark (1995), *Lewis Mumford and the Ecological Region*, The Guilford Press, New York.

MacKaye, Benton (1962), *The New Exploration: A Philosophy of Regional Planning*, University of Illinois Press, Urbana.

Index

For Product Safety Concerns and Information please contact our EU representative GPSR@taylorandfrancis.com Taylor & Francis Verlag GmbH, Kaufingerstraße 24, 80331 München, Germany

Batch number: 10726007

Printed by Printforce, the Netherlands